NEOTROPICAL PRIMATES
Field Studies and Conservation

Proceedings of a Symposium on the
DISTRIBUTION AND ABUNDANCE OF NEOTROPICAL PRIMATES

R. W. THORINGTON, JR.
P. G. HELTNE
Editors

COMMITTEE ON CONSERVATION OF NONHUMAN PRIMATES
Institute of Laboratory Animal Resources
Assembly of Life Sciences
National Research Council

NATIONAL ACADEMY OF SCIENCES
Washington, D.C.   1976

NOTICE: The project that is the subject of this report was approved by the Governing Board of the National Research Council, whose members are drawn from the Councils of the National Academy of Sciences, the National Academy of Engineering, and the Institute of Medicine. The members of the Committee responsible for the report were chosen for their special competences and with regard for appropriate balance.

This report has been reviewed by a group other than the authors according to procedures approved by a Report Review Committee consisting of members of the National Academy of Sciences, the National Academy of Engineering, and the Institute of Medicine.

This symposium was sponsored by the Battelle Seattle Research Center and the Institute of Laboratory Animal Resources, National Research Council. It was supported in part by Contract PH43-64-44 (National Institutes of Health, Public Health Service), Contract AT (11-1)-3369 (Atomic Energy Commission), Contract N00014-67-A-0244-0016 (Office of Naval Research, U.S. Army Medical Research and Development Command, and U.S. Air Force), Contract 12-16-140-155-91 (U.S. Department of Agriculture), Contract NSF-C310 (National Science Foundation); Grant RC-10 (American Cancer Society), and contributions from pharmaceutical companies and other industry.

Library of Congress Cataloging in Publication Data
Main entry under title:

Neotropical primates.

"Sponsored by the Battelle Seattle Research Center and the Institute of Laboratory Animal Resources, National Research Council."
Includes bibliographies and index.
1. Primates—Congresses. 2. Mammal populations—Latin America—Congresses. I. Thorington, Richard W. II. Heltne, Paul G. III. Battelle Memorial Institute. Columbus, Ohio. Seattle Research Center. IV. National Research Council. Institute of Laboratory Animal Resources.
QL737.P9N38   599'.8'098   76-10786
ISBN 0-309-02442-0

Available from

Printing and Publishing Office
National Academy of Sciences
2101 Constitution Avenue, N.W.
Washington, D.C. 20418

Printed in the United States of America

80 79 78 77 76   10 9 8 7 6 5 4 3 2 1

PREFACE

During the last decade there has been increasing concern over the status of wild populations of primates, a concern expressed by conservationists and by many persons involved in medical research. In response, the Institute of Laboratory Animal Resources (ILAR) established the Committee on Conservation of Nonhuman Primates, which has sought information on the many facets of the problem and has developed certain strategies for the future. As one measure, a symposium on the distribution and abundance of neotropical primates was held in August 1972 at the Battelle Conference Center, in Seattle, Washington.

There have been significant additions to our knowledge of the distribution and abundance of neotropical primates in the period since the symposium was held and the papers were prepared. It is gratifying that many such contributions have come from persons who attended or contributed directly to the symposium. We are confident that the material presented here is significant and provides essential documentation of progress in the study of platyrrhine biology.

We are grateful to Drs. Thorington and Heltne for their editorial efforts and to the following for talented assistance at various stages in preparing the proceedings for publication: Dr. Miguel A. Schön, Linda LeResche, Robert Johnson, Johanne Humphrey, and the ILAR staff—especially Dr. Nancy A. Muckenhirn and Lydia E. Koutzé. We are especially thankful to Dr. Charles Southwick, present chairman of the ILAR Committee on Conservation of Nonhuman Primates, who has been most helpful in bringing this symposium to publication.

COMMITTEE ON CONSERVATION OF NONHUMAN PRIMATES (1972)

Richard W. Thorington, *Chairman*
Stuart A. Altman
Clarence R. Carpenter
Joel E. Cohen
Arnold F. Kaufmann
Donald G. Lindburg
Charles H. Southwick
John G. Vandenbergh

CONTENTS

INTRODUCTION

Richard W. Thorington, Jr., *and* Paul G. Heltne

New World monkeys have constituted about one-quarter of all primates utilized for biomedical purposes in recent years in the United States. Next to the rhesus monkey (*Macaca mulatta*), the South American squirrel monkey (*Saimiri sciureus*) has been the most frequently used experimental primate. Several of the New World species are critical for certain types of research, e.g., the night monkey (*Aotus*) for studies of malaria (Millar, 1974) and some of the tamarins of the genus *Saguinus* for studies of viral sarcoma and hepatitis (Deinhardt, 1971). The ceboid monkeys offer special opportunities for the study of the ecology of coexistent primate species in highly diverse communities. Some species, like the marmosets, provide excellent subjects for investigation of parental behavior. Recent work has pointed to the extreme similarities between the morphology of some extant New World monkeys and *Proconsul africanus* of the African Miocene (Schön and Ziemer, 1973). Other anatomical studies have emphasized similarities of hip morphology in certain ceboids and man (Stern, 1971). So little is known of most New World genera that further research will almost certainly pay unexpected dividends for medicine and biology.

Not widely appreciated is the fact that these important species are under severe pressure due to the widespread destruction of their forest habitats. In an effort to emphasize the present status of primate populations, the organizers of this symposium invited participants who had done recent fieldwork in Central and South America. They were requested to summarize available population data and ecological information relevant to the present and future conditions of wild neotropical primates. There were obvious gaps in coverage, for example, a lack of papers on Brazilian primates. To our knowledge, there are simply no data available on the status of Brazilian populations of primates in the Amazon. Certain taxonomic underrepresentations were also evident, such as the absence of material dealing with any of the species of the Pithecinae, whereas studies of *Alouatta* are overrepresented. We believe these omissions and overrepresentations, both geographically and taxonomically, were a valid reflection of the information, and lack of information, on South American monkeys at the time of the symposium. The situation has changed only slightly since then. We hope that publication of this volume and the recognition of hiatuses will stimulate future work that will narrow these gaps.

The distribution of primates is usually documented from museum collections. This has the advantage that the identifications can be reevaluated and verified. There are several disadvantages. Most collections are old and specimens may have been obtained from areas in which the species no longer occurs due to habitat changes or other factors. Also, museum collections never document the ranges of animals as thoroughly as the local naturalist can. Nevertheless, good descriptions of the geographic ranges of South American primates are available in the literature (e.g., Lawrence, 1933; Kellogg and Goldman, 1944; Hershkovitz, 1949, 1963, 1968; Cabrera, 1957; Fooden, 1963). Unfortunately, the most readily available syntheses (Hill, 1957, 1960, 1962) of geographic ranges, complete with

1

maps, contain a number of errors and misinterpretations of the earlier literature. The chapter in this volume by Hernández-Camacho and Cooper is a synthesis of museum work, literature survey, and extensive firsthand knowledge of the areas treated. It is thus a valuable addition to the documentation of geographic ranges. Other papers in this symposium document the occurrence of primates in particular places but do not take as broad a view of the distribution questions as do Hernández-Camacho and Cooper of Colombia. There is a great need for this type of precise and current documentation of the ranges of New World monkeys. The need for continual monitoring of the ranges is evident when we consider that wild primates no longer occur at the study sites utilized by the Baldwins and the Kleins. Recent work in northern Colombia (Struhsaker *et al.*, 1975) has also documented that primate ranges there have shrunk significantly from those reported by Hernández-Camacho and Cooper.

Distribution maps in the literature seldom convey information about the discontinuous distribution of animals in nature. The habitat preferences of neotropical primates are not well documented, nor are the population densities that occur in different habitats. The data from Barro Colorado Island (BCI) (Carpenter, 1934; Altmann, 1959; Oppenheimer, 1968; Chivers, 1969; Hladik and Hladik, 1969; Richard, 1970; and many others) have provided some documentation for *Alouatta palliata, Cebus capucinus, Saguinus oedipus, Ateles geoffroyi,* and *Aotus trivirgatus.* Ecological data have been provided on other species by Mason (1966), Thorington (1967, 1968), Baldwin and Baldwin (1971), and a number of others. Most of the chapters in this volume add additional information about habitat utilization. However, we are just beginning to obtain the comparative data that will enable us to define precisely the different ways in which various habitats are used by primates who share them. The Kleins have provided us with a good model of this type of study. In future studies we will need to account much more carefully for the spatial distribution of resources and how the distribution of available food changes seasonally. In the neotropics this requires knowledge of the several hundred species of vascular plants per hectare that are available to the herbivorous and frugivorous species of monkeys. The insectivorous primates, like *Saimiri, Aotus,* and the marmosets provide even more difficult problems, and it may be a long time before we can describe their utilization of habitats more precisely than by describing their foraging habits.

Ultimately, analyses of habitat utilization and evolutionary success require knowledge of the population densities of the animals. Although much has been written on the subject of censusing (Thorington, 1972; Eisenberg and Thorington, 1973), the state of the art is still not very satisfactory. Some of the problems and solutions to censusing different species of primates are considered in the chapters by Freese, the Baldwins, the Kleins, Neville, and Heltne *et al.* The problems were discussed further at the symposium, but it is obvious that there is much still to be resolved. The major problems continue to be the conversion of observational data to absolute densities; the testing and demonstration of consistency among observers, techniques, and habitats; and the demonstration of accuracy.

The problems of determining population structure also received a great deal of attention, particularly in the chapter by Heltne *et al.* and the subsequent discussions. There are ways to estimate the ages of monkeys in the wild or in the hand (Carpenter, 1934, for *Alouatta;* Long and Cooper, 1968, for *Saimiri;* Tappen and Severson, 1971, for *Saimiri, Cebus,* and *Saguinus;* and Thorington and Vorek, in press, for *Aotus*). These need to be extended to other genera; tested against known-age animals, both in the lab and in the field; and adapted and refined more for fieldwork. Again the accuracy and interobserver reliability need to be scrutinized carefully.

The medical community in the United States is making efforts to establish breeding colonies of neotropical primates to supply the most essential research needs for these animals. It is conceivable that a series of such colonies in South American countries and in the United States could make medical research less dependent on wild populations of these monkeys. However, the basic problems of conservation still exist. The human population of Latin America continues to increase at a rapid rate, and the aspirations of the governments and the people for economic development are understandably increasing at an even more rapid rate. These facts are reflected in the rate of destruction of forest and conversion of the land into agricultural or pastoral production. Brazil is reputed to have converted 84 million acres into farmland and to have logged an additional 26 million acres in the last decade. A map of northern Colombia, prepared in 1966 to show the distribution of forests, bears no resemblance to the pictures obtained from satellites in 1973. Most of the forests have been completely cut down. These examples are indicative of the changes that are occurring throughout South America. There is an immediate need to establish and protect more reserves in which neotropical primates can survive. There is an increasing need for more data on the ecology and population dynamics of these animals to permit educated management of reserves and parks in which they occur.

REFERENCES

Altmann, S. A. 1959. Field observations on a howling monkey society. J. Mammal. 40:317–330.

Baldwin, J. D., and J. I. Baldwin. 1971. Squirrel monkeys (*Saimiri*) in natural habitats in Panama, Colombia, Brazil and Peru. Primates 12:45–61.

Cabrera, A. 1957. Catalogo de los mamiferos de America del Sur. Cienc. Zoolog. IV(1):133–202.

Carpenter, C. R. 1934. A field study of the behavior and social relations of howling monkeys. Comp. Psychol. Monogr. 10(2):1–168.

Chivers, D. J. 1969. On the daily behavior and spacing of howling monkey groups. Folia Primatol. 10:48–102.

Deinhardt, F. 1971. Use of marmosets in biomedical research. Pages 918–925 *in* E. I. Goldsmith and J. Moor-Jankowski, eds. Medical primatology 1970. S. Karger, Basel, Switzerland.

Eisenberg, J. F., and R. W. Thorington, Jr. 1973. A preliminary analysis of a neotropical mammal fauna. Biotropica 5:150–161.

Fooden, J. 1963. A revision of the woolly monkeys (genus *Lagothrix*). J. Mammal. 44:213–247.

Hershkovitz, P. 1949. Mammals of northern Colombia. Preliminary report no. 4: Monkeys (Primates), with taxonomic revision of some forms. Proc. U.S. Nat. Mus. 98:323–427.

Hershkovitz, P. 1963. A systematic and zoogeographic account of the monkeys of the genus *Callicebus* (Cebidae) of the Amazonas and Orinoco River basins. Mammalia 27:1–79.

Hershkovitz, P. 1968. Metachromism or the principle of evolutionary change in mammalian tegumentary colors. Evolution 22:556–575.

Hill, W. C. O. 1957. Primates. III. Pithecoidea. Interscience Publishers, Inc., New York.

Hill, W. C. O. 1960. Primates. IV. Cebidae, Part A. Interscience Publishers, Inc., New York.

Hill, W. C. O. 1962. Primates. V. Cebidae, Part B. Interscience Publishers, Inc., New York.

Hladik, A., and C. M. Hladik. 1969. Rapports trophiques entre végétation et primates dans la forêt de Barro Colorado (Panama). La Terre et la Vie 23:25–117.

Kellogg, R., and E. A. Goldman. 1944. Review of the spider monkeys. Proc. U.S. Nat. Mus. 96:1–45.

Lawrence, B. 1933. Howler monkeys of the *palliata* group. Bull. Mus. Comp. Zool. 75(8):315–354.

Long, J. O., and R. W. Cooper. 1968. Physical growth and dental eruption in captive-bred squirrel monkeys, *Saimiri sciureus* (Leticia, Colombia). Pages 193–205 *in* L. A. Rosenblum and R. W. Cooper, eds. The squirrel monkey. Academic Press, New York.

Mason, W. A. 1966. Social organization of the South American monkey *Callicebus moloch:* a preliminary report. Tulane Stud. Zool. 13:23–28.

Millar, J. W. 1974. Malaria. Pages 13–18 *in* Forty-fifth annual report of the work and operations of the Gorgas Memorial Laboratory, Fiscal Year 1973. U.S. Government Printing Office, Washington, D.C.

Oppenheimer, J. R. 1968. Behavior and ecology of the white-faced monkey, *Cebus capucinus,* on Barro Colorado Island, C.Z. Doctoral Dissertation. University of Illinois.

Richard, A. 1970. A comparative study of the activity patterns and behavior of *Alouatta villosa* and *Ateles geoffroyi*. Folia Primatol. 12:241–263.

Schön, M. A., and L. K. Ziemer. 1973. Wrist mechanism and locomotor behavior of *Dryopithecus (Proconsul) africanus*. Folia Primatol. 20:1–11.

Stern, J. T., Jr. 1971. Functional myology of the hip and thigh of ceboid monkeys and its implications for the evolution of erect posture. Bibl. Primatol. no. 14. Karger, Basel, Switzerland.

Struhsaker, T. T., K. Glander, H. Chirivi, and N. J. Scott. 1975. A survey of primates and their habitats in northern Colombia. Pages 43–78 *in* Primate censusing studies in Peru and Colombia. Pan American Health Organization, Washington, D.C.

Tappen, N. C., and A. Severson. 1971. Sequence of eruption of permanent teeth and epiphyseal union in New World monkeys. Folia Primatol. 15:293–312.

Thorington, R. W., Jr. 1967. Feeding and activity of *Cebus* and *Saimiri* in a Colombian forest. Pages 180–184 *in* R. Schneider, H. J. Kuhn, and D. Starck, eds. Neue ergebnisse de primatologie. Fischer, Stuttgart.

Thorington, R. W., Jr. 1968. Observations of squirrel monkeys in a Colombian forest. Pages 69–85 *in* L. A. Rosenblum and R. W. Cooper, eds. The squirrel monkey. Academic Press, New York.

Thorington, R. W., Jr. 1972. Censusing wild populations of South American monkeys. Pages 26–32 *in* International movement of animals. Pan American Health Organization, Washington, D.C.

Thorington, R. W., Jr., and R. E. Vorek. In press. Observations on the geographic variation and skeletal development of *Aotus*. Lab. Anim. Sci.

CENSUSING *ALOUATTA PALLIATA*, *ATELES GEOFFROYI*, AND *CEBUS CAPUCINUS* IN THE COSTA RICAN DRY FOREST

Curtis Freese

INTRODUCTION

This chapter describes the distributions, relative population levels, and habitat preferences of the howler monkey (*Alouatta palliata*), spider monkey (*Ateles geoffroyi*), and white-faced monkey (*Cebus capucinus*) in Santa Rosa National Park, located in the tropical dry forest along the Pacific coast of northwestern Costa Rica. Data on troop size and composition, territory size, and reproduction also are presented.

DESCRIPTION OF STUDY AREA

Santa Rosa is approximately 100 km² in size and lies approximately 30 km northwest of Liberia, in Guanacaste Province, with the Santa Elena Peninsula bordering on the northwest and the Inter-American Highway on the northeast (Figure 1). The topography of the park varies from nearly flat to steep, impassable canyon walls and hillsides. Elevation ranges from sea level to over 350 m, with most of the park on a plateau at approximately 300 m. Typical of the tropical dry forest in Costa Rica, the park is a mosaic of grasslands and woodlands. The stands of trees vary through secondary-growth stages to nearly mature deciduous and evergreen forest. By way of definition, in areas of deciduous forest nearly all trees lose their leaves during the dry season. In semideciduous forest areas, roughly one-half of the trees lose their leaves during the dry season. In areas of evergreen forest, nearly all of the trees maintain their leaves year-round.

Santa Rosa receives approximately 1,750 mm of rain annually, mostly from June to November. During the pronounced dry season, from December to May, the deciduous forests, which dominate most areas in the park, lose their leaves.

The canopy heights of deciduous forests in the park range from about 5 to 15 m, semideciduous from about 10 to 20 m, and evergreen from about 15 to 40 m. The divisions between the three main types, particularly between deciduous and semideciduous, are somewhat arbitrary since continuous gradations exist. Also, it is difficult to decide how one should define a primarily deciduous forest containing large islands of evergreen trees. Studies of succession, species composition, productivity, and phenology of flowering, fruiting, and leafing in the forests of Santa Rosa are needed to illuminate differences between the forest types.

METHODS

The data in this chapter were obtained from a study conducted intermittently from October 1971 to April 1972. Most of the parkwide census was conducted in the dry season months of January, February, and March. Generally, we collected data by recording observations of monkey troops as we walked along trails, in stream valleys, and through trailless forests. The morning howls of howler monkeys assisted in locating them. When a monkey troop was encountered, the location and forest type were noted, the number in the troop counted or estimated, and, when possible, individuals were identified as to sex and age.

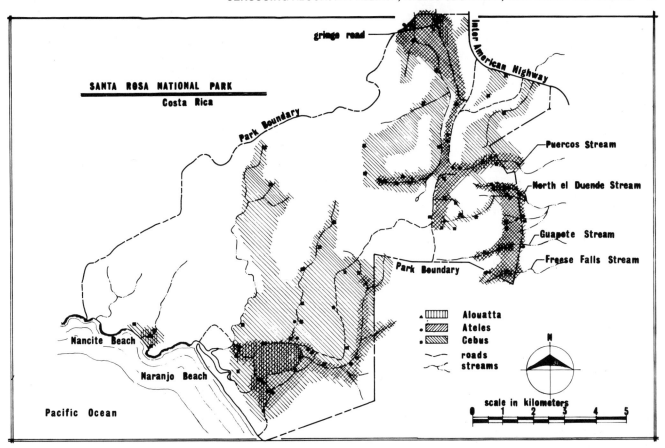

FIGURE 1 Locations and distribution of *Alouatta, Ateles,* and *Cebus* sightings in Santa Rosa National Park. Only those sightings important in outlining species ranges are plotted. The approximate range sizes as outlined for each species are: *Alouatta*—4 km², *Ateles*—17 km², *Cebus*—49 km².

RESULTS

General Distribution and Forest Preferences

The approximate distributions of each of the three species (Figure 1) were extrapolated from the distribution of the forest types in which groups of the species were sighted. The following observations are at variance with Moynihan's (1970) statement that the three species are ". . . largely or completely inhabitants of tall forests."

During the dry season *Alouatta palliata* appear to utilize, almost exclusively, areas of predominantly mature evergreen forest. In 50 percent of the cases where troops were found, the evergreen forest was strictly riparian, with deciduous to semideciduous forest dominating the surrounding area. Some of the riparian evergreen forests are small, particularly those occurring along streams crossing the eastern boundary of the park. The preference of *Alouatta* for evergreen forest corroborates Carpenter's (1934) conclusion that

this species usually avoids scrub forest, preferring instead primary forest. *Alouatta* apparently are not expanding into or utilizing certain evergreen forests contiguous with areas they inhabit. It is possible, however, that *Alouatta* move into deciduous forests areas to feed during the wet season when the trees are in leaf. Indeed, even during the dry season, *Alouatta* have been seen or heard in deciduous forests in other areas of the Costa Rican dry forest (R. Carroll, K. Leber, and D. Boucher, personal communication). *Alouatta* have also been reported to live in tall, dry, deciduous forests about 75 km southeast of Santa Rosa (Glander, 1971). Consequently, their distribution may be somewhat larger than delineated in Figure 1.

Ateles geoffroyi seem to frequent both evergreen and semideciduous forests during the dry season. At times they may even visit a deciduous forest. Due to this wider utilization of forest types, their distribution in Santa Rosa is more extensive than that of *Alouatta*. It is possible, particularly during the wet season, that *Ateles* also range over a more extensive area than outlined in Figure 1. However, *Ateles* appear to center most of their dry season activity in the evergreen forests and small, isolated islands of evergreen trees in semideciduous forests.

The differential distribution of *Alouatta* and *Ateles* in Santa Rosa supports Richard's (1970) observations on Barro Colorado Island. She describes *Ateles* on Barro Colorado Island as living in ". . . small, highly mobile groups . . . ," which ". . . range over comparatively wide areas to secure a catholic diet with a relatively high energy content." *Alouatta,* by contrast, ". . . lives in medium-sized groups which move slowly between food sources, which they exploit until it is no longer economic to do so."

Cebus capucinus are the most widespread monkeys in the park and are seen in every forest type. One troop was encountered in the mangrove forest near Naranjo Beach. *Cebus* will also move out into very sparsely forested areas by following a long, nearly single line of short, deciduous trees in order to reach a fruiting tree. In addition, individuals have been seen on the ground several times (e.g., a group of four or five moved about 100 m over open ground from one part of a forest to another; they apparently were avoiding the tree route that would have led them directly past humans). On Barro Colorado Island, *Cebus* apparently spent a great deal of time in the lower levels of tall trees as well as frequenting shorter trees (Collias and Southwick, 1952; Oppenheimer, 1968). Thus, the utilization by *Cebus* of short, secondary forests in Santa Rosa is not exceptional.

Population Size, Troop Size and Composition, Territory Size, and Reproduction

The success of securing data on population size, troop size and composition, territory size, and reproduction varied a great deal among species. For all three species, only young that were clinging to the mother during movement were classified as infants; all other animals of less than subadult size were classified as juveniles.

Table 1 shows the size and composition of *Alouatta* troops in Santa Rosa. The total *Alouatta* population in the park is probably between 70 and 100 individuals distributed in 8 to 10 troops. The three largest troops occur in rather extensive stands of evergreen forest, none of which are strictly riparian. The three smaller troops of the Freese Falls Stream, Guapote Stream, and North El Duende Stream, representing complete or almost-complete counts, inhabit small stands of evergreen forest along streams. The limited area of these evergreen forests may be a major determinant of the small numbers in these troops. There were indica-

TABLE 1   Composition of *Alouatta* Troops

Location	Adult Males	Adult Females	Juveniles	Infants	Total
Naranjo Basin	4	10	7	3	24
Naranjo Basin	1	5	3	3	12[a]
Gringo Road	1	4	3	1	9
Puercos Stream	1	4	0	0	5[a]
Freese Falls Stream	2	1	2	0	5[a]
Guapote Stream	1	2	1	0	4
North El Duende Stream	2	1	0	0	3[a]
Nancite Basin	1	2	0	0	3[a]

[a]Possibly an incomplete count.

tions that one or two other large troops inhabit the evergreen forest of Naranjo Beach, where the two largest troops were found.

Adult males comprised approximately 20 percent of the censused population; adult females, 44 percent; juveniles, 24 percent; and infants, 12 percent. Previous investigators likewise have found a much higher proportion of adult females that adult males in *A. palliata*, although the ratio has been variable (Carpenter, 1934; Collias and Southwick, 1952; Chivers, 1969; Glander, 1971).

As seen in Table 1, the large troops appear to have the highest ratio of young to adults. Infants and juveniles composed about 46 percent of the three largest troops but only about 13 percent of the five smallest troops. The sample size is very small, but these findings agree with those of Collias and Southwick (1952), which indicate that in *Alouatta* ". . . disproportionately small numbers of young may be produced in clans of very restricted size. . . ." Glander (1971) also notes a disproportionately small number of young in the two smallest *Alouatta* troops he censused in the dry forest. Comparing the percentages of young in Table 1 with figures presented by earlier investigators (Carpenter, 1937; Collias and Southwick, 1952), it seems probable that the reproductive rate in the larger troops is sufficient for maintaining their present numbers, though in smaller troops the reproductive rate may not be adequate for replacement.

Newly born infants were observed during the wet and dry seasons, although the data are insufficient to determine the occurrence of a seasonal peak in births. Glander (1971) found evidence of a birth season for *Alouatta* at his study site, which is also in the tropical dry forest of Guanacaste Province, Costa Rica.

Data on size and composition of *Ateles* groups are presented in Tables 2 and 3. Single individuals were also counted as a group. Only complete group-size

TABLE 2   Composition of Completely Censused *Ateles* Groups Given in the Order They Were Seen

Adult Males	Adult Females	Juveniles	Infants	Total
2				2
1				1
3				3
	3			3
	3			3
3				3
2				2
	2			2
1	2			3
	1			1
1				1
	3	1	1	5
2				2
1				1
2	6		2	10
2	3	3	1	9
1	5	2–3	3	11–12
2	4	2	2–3	10–11
	2			2
	1	1		2
	1	1		2
2	1	1	1	5

counts were used in Table 2, thus biasing these data toward the more easily censused small groups. Also, there would be a slight shift in the age–class frequencies if some subadults would have been classified as juveniles instead of as adults.

As a rough estimate, the *Ateles* population in Santa Rosa is between 110 and 160 individuals. Eighty-three individuals were observed in 22 complete counts (some individuals were likely counted more than once). Of these, 30 percent were adult males, 45

TABLE 3   Group Size Frequency in *Ateles*

Group Size[a]	Frequency
1	7
2	12
3	6
4	2
5	7
6	2
8	2
9	1
10	3
11	1
12	1
15–20	1

[a]Some estimated.

percent adult females, 14 percent juveniles, and 11 percent infants. These findings agree with those of Carpenter (1935), who also found that *Ateles* populations in Panama contain more adult females than adult males. However, Carpenter found that most of his subgroups contained adults of both sexes, whereas 16 of the 22 completely censused groups in this study contained adults of only one sex. Table 3 shows that group size ranged from 1 to 20 or more, with an average of about 4.5, smaller than Carpenter's average group size appeared to be. Group sizes of 1–3 and 5 are frequent with a group of two individuals being the most frequent size observed.

Apparently the social structure of *Ateles* varies with the habitat (Eisenberg and Kuehn, 1966). While Carpenter (1935) reported that *Ateles* live in large troops typically composed of smaller subgroups in a tall, evergreen forest of Panama, Eisenberg and Kuehn (1966) found that this species lives in small, cohesive groups in a mangrove swamp in Mexico. Conditions in Santa Rosa during the dry season probably more closely approach the marginal conditions of the mangrove swamp than those of the tall forest in Panama. The social structure of *Ateles* groups in Santa Rosa at times resembles that found by Eisenberg and Kuehn in Mexico. At other times the structure resembles that observed by Carpenter in Panama. For example, I twice followed a group of three females (probably the same group both times) to a sleeping tree, where they remained alone during the night. On other occasions, I have watched small groups moving in close proximity or even joining together, indicating that large groups sometimes are formed by the union of smaller groups, as found by Carpenter (1935). It is possible that the grouping tendencies of *Ateles* within the park vary between groups inhabiting semideciduous to deciduous forests and those inhabiting extensive evergreen forests. The grouping tendency may also vary between the wet and dry seasons.

Carpenter (1935) concluded that "spider monkeys in the wild have no distinct breeding season," although there may be considerable fluctuations in the number of births during different times of the year. At Santa Rosa a higher proportion of infants to total *Ateles* was noted during the middle of the dry season. This peak, however, lasted only two months, whereas infants are carried much longer than that.

Cebus capucinus are difficult to census, due to the large size of the troops and the members' proclivities for dispersion and continual movement. Rough estimates of troop size were often impossible to attain, and the census of any *Cebus* troop allowed for a certain margin of error. I estimated that there were between 15 and 20 *Cebus* troops in Santa Rosa and that the park's population was between 250 and 350 individuals.

Most troops contained between 15 and 20 individuals. Infrequently, small groups of only one to six individuals were encountered, sometimes under circumstances indicating that particular groups were temporarily separated from larger troops. Adult females outnumbered adult males. One troop inhabiting a semideciduous forest contained about five adult males (one was subadult), six or seven adult females (at least two were subadults), three or four juveniles, and two infants. If individuals classified as subadults were included instead in the juvenile category, the age distribution for this troop would more closely approximate Oppenheimer's (1968) results.

The largest troop size observed by Oppenheimer (1968) on Barro Colorado Island was 15, although he noted that significantly larger troops have been seen by other observers. He also found there were more females than males and usually twice as many infants and juveniles as adults. The territory of one Santa Rosa troop was estimated at approximately 0.5 km², as compared to an average of about 0.9 km² for slightly smaller troops observed by Oppenheimer. Oppenheimer's data suggest that more *Cebus* births occur during the dry season or early wet season on Barro Colorado Island. I observed infants during every month and noted one birth at the beginning of the dry season and another at the beginning of the wet season. However, these data are too inconclusive to determine the presence or absence of a seasonal reproductive peak for *Cebus* in Santa Rosa.

SUMMARY

Of the three monkey species in Santa Rosa National Park in the dry forest of Guanacaste Province, Costa Rica, *Cebus* were the most numerous, *Ateles* intermediate in number, and *Alouatta* the least numerous. Similarly, *Cebus* had the most extensive distribution and *Alouatta* the smallest. John Christy (personal communication) has found in another area of the Costa Rican dry forest that *Alouatta* are more numerous than *Cebus* and that *Ateles* are absent.

It seems likely that *Alouatta* inhabit almost exclusively evergreen forests in the park because they are largely folivores. During the dry season leaves are scarce in the deciduous forests, and the howlers are not mobile enough to regularly penetrate deciduous forests to reach fruiting trees. *Ateles* probably have a larger proportion of fruits to leaves in their diet than *Alouatta,* and they are much more mobile. Because dry-season fruiting is common in the deciduous forest, *Ateles* can obtain the most important part of their diet

in the deciduous forest, as well as in the evergreen forests, year-round. And they can easily cover long distances to obtain fruit in deciduous forests and leaves in evergreen forests. Also, the small groups of *Ateles* can probably more efficiently utilize the widely scattered fruit sources than the large *Alouatta* troops, because more time can be spent at each source before it is exhausted. Similarly, the differences in the diet and locomotor abilities of *Cebus* compared to the other two species suggest why they are more widespread in Santa Rosa. *Cebus* eat primarily fruits and insects, both of which occur year-round in varying abundance in the deciduous forests. Insects, but possibly not fruits, also can be found year-round in young, secondary, deciduous growth; and insect-foraging by *Cebus* appears to be a frequent activity in such growth. *Cebus* are usually widespread during insect-foraging so that each individual is a distinct foraging unit covering a different area. Important for exploiting young, secondary growth are the comparatively small size of *Cebus* and their quadrupedal locomotion, which allow easier movement through the weakly structured, dense, short vegetation. The physiognomy and types of food resources of young, deciduous forests may largely exclude their utilization by *Ateles* and *Alouatta*.

The extremely variable forests of Santa Rosa contrast with the relatively uniform, tall, evergreen forest on Barro Colorado Island. Within a small area, apparent differences in suitable forest habitat exist for each species. Models of the division of resource space can be tested. The marked seasonality and relative lack of uniformity of the tropical dry forest, as compared to the tropical wet forest, have certainly imposed different pressures upon resident primates. These pressures may be reflected in behavioral, physiological, and other biological differences between tropical wet and tropical dry forest monkey populations. Comparative studies of troops of the same species in different habitats within the park and between wet and dry seasons could greatly enhance our understanding of socioecology in New World primates. Kummer (1971) has discussed the importance of understanding ecological adaptation and resource partitioning among primates.

Beyond the boundaries of the park, the rapid cutting and clearing of the tropical dry forest in northwestern Costa Rica continues. *Ateles* are already scarce in this region, and *Alouatta* and *Cebus* populations will be seriously affected if present trends continue. If sizeable populations of the primate species and other fauna are to persist, land-use practices must change. Hopefully, the continued protection of wildlife and forests in Santa Rosa by the Costa Rican government will insure their survival under more or less natural conditions in this biologically unique area.

ACKNOWLEDGMENTS

Special thanks are due to the other three biologists in the park—Douglas Boucher, Steve Cornelius, and Keither Leber—whose observations and helpful suggestions have added immensely to the content of this chapter. I also want to thank the director of the Costa Rica National Parks Department, Mario Boza, for his cooperation, and the administrator of Santa Rosa, Alvaro Ugalde, for his cooperation and valuable observations. I extend my appreciation to Dr. Paul Heltne, who offered helpful suggestions on the manuscript. Finally, my gratitude goes to the park guards, who often pointed the way to a monkey troop.

REFERENCES

Carpenter, C. R. 1934. A field study of the behavior and social relations of howling monkeys (*Alouatta palliata*). Comp. Psychol. Monogr., 10:1–68.

Carpenter, C. R. 1935. Behavior of red spider monkeys in Panama. J. Mammal. 16(3):171–180.

Chivers, D. J. 1969. On the daily behavior and spacing of howling monkey groups. Folia Primatol., 10:48–102.

Collias, N. E., and C. H. Southwick. 1952. A field study of population density and social organization in howling monkeys. Proc. Am. Philos. Soc., 96:143–146.

Eisenberg, J. F., and R. E. Kuehn. 1966. The behavior of *Ateles geoffroyi* and related species. Smithson. Misc. Collect., 151(8):1–63.

Glander, K. 1971. The howlers of Finca La Pacifica. M.S. Thesis. Univ. of Chicago. 90 pp.

Kummer, H. 1971. *Primate Societies*. Aldine and Atherton, Chicago and New York. 160 pp.

Moynihan, M. 1970. Some behavior patterns of platyrrine monkeys. II. *Saguinus geoffroyi:* and some other tamarins. Smithson. Contrib. Zool., no. 28.

Oppenheimer, J. R. 1968. Behavior and ecology of the white-faced monkey, *Cebus capucinus*. Ph.D. Thesis. Univ. of Illinois. 181 pp.

Richard, A. 1970. A comparative study of the activity patterns and behavior of *Alouatta villosa* and *Ateles geoffroyi*. Folia Primatol., 12:241–263.

COMPARISON OF CENSUS DATA ON *ALOUATTA PALLIATA* FROM COSTA RICA AND PANAMA

Paul G. Heltne, Dennis C. Turner, *and* Norman J. Scott, Jr.

INTRODUCTION

This chapter evaluates the status and trends of two populations of the mantled howler monkey (*Alouatta palliata* following the usage of Smith, 1970; *A. villosa* of Hall and Kelson, 1959). Between 1966 and 1971, graduate field teams of the Organization for Tropical Studies (OTS) conducted 12 censuses of howler monkeys in the tropical dry forest of northwestern Costa Rica. The Costa Rican data are compared to the censuses of the same species of howler living on Barro Colorado Island (BCI), Panama, a nature reserve less harshly seasonal than the Costa Rican sites and with no serious alterations since the completion of the Panama Canal.

REVIEW OF PREVIOUS CRITERIA

This section will review previous criteria and suggest a new indicator of the stability of a troop or population. Several statistics have been used to signal whether a population is in a depressed, stable, or increasing condition. When an exhaustive census of an isolated area is conducted, the total number of monkeys is obviously a key factor to compare with former tallies. Collias and Southwick (1952) counted a drastically reduced howler population following a yellow fever epidemic on BCI, and statistics derived from their census are regarded as signs of a depressed population; namely, low average troop size and an increased number of females per male. They reported an average troop size of 7.9 for 1951 (Table 3); later troop size

averages were more than twice this figure (Carpenter, 1964). The adult male: adult female (M:F) ratio was 1:3.75. The Collias and Southwick study also suggested that a low proportion of females with infants is indicative of a population in difficulty. All of these criteria are further advanced by Carpenter (1964).

In addition, the ratio of adult females to the sum of juveniles plus infants (F:J + I) may be a useful measure of whether a troop or population is maintaining, increasing, or decreasing its numbers. Tables 1 and 2 reveal the reason for attention to the F:J + I ratio. The illustrations presented in these two tables are based on assigning values to certain life history variables, namely: (a) number of safe reproductive years for the adult female, (b) number of young per year per female (all the young born in one year compose a cohort), (c) sex ratio at birth, (d) mortality before completion of breeding potential in each birth cohort, (e) relative effect of mortality on the sexes, (f) times at which mortality is assessed, and (g) number of years to maturity.

In Part I of Table 1, the values assigned to the above categories are as follows: (a) the adult female has 4 safe reproductive years, (b) one infant is born per year per female, (c) the sex ratio at birth is 1:1, (d) there is a 50 percent preadult mortality of each new cohort, (e) both sexes are affected equally by the mortality, (f) the mortality is assessed at birth, and (g) 3 years intervene between birth and maturity. These particular assumptions will subsequently be referred to as the "7-year model." Any troop or population with such characteristics should show an F:J + I ratio of around 1:1.5 if it is to remain stable. Stability, here, is defined as maintaining a steady supply of adult females. Females with dependent infants (those in the 0–1 age group) would comprise about 50 percent of the total adult female population; the F:I ratio would be about 1:0.50.

TABLE 1 Models for Stable or Increasing Populations

Models for 4 years of active reproduction for females reaching maturity after 3 preadult years; 1 infant/year/female; 1:1 sex ratio among infants; no sex differential in preadult mortality. † indicates occurrence of a death.

Part I—7-Year Model
Number of females is constant; local population size constant; all preadult mortality assessed at birth.

Age in Years							
I 0–1	J 1–2	J 2–3	F 3–4	F 4–5	F 5–6	F 6–7	F 7–8
Year 1 M_0 F_0 $M^\dagger$ $F^\dagger$	M_1 F_1	M_2 F_2	F_3	F_4	F_5	F_6	$F_7^\dagger$
Year 2 M F $M^\dagger$ $F^\dagger$	M_0 F_0	M_1 F_1	F_2	F_3	F_4	F_5	$F_6^\dagger$
			F:J + I = 1:1.5				

Average troop size given four adult females per troop: 2I + 4J + 4F + 1M (or more) = 11 (or more).

Part II—Alternate 7-Year Model
Number of females is constant; local population size constant; allocation of preadult mortality extended over 2 years.

Age in Years							
I 0–1	J 1–2	J 2–3	F 3–4	F 4–5	F 5–6	F 6–7	F 7–8
Year 1 M_0 F_0 M_0 $F_0^\dagger$	M_1 F_1 $M_1^\dagger$	M_2 F_2	F_3	F_4	F_5	F_6	$F_7^\dagger$
Year 2 M F M $F^\dagger$	M_0 F_0 $M_0^\dagger$	M_1 F_1	F_2	F_3	F_4	F_5	$F_6^\dagger$
			F:J + I = 1:1.75				

Average troop size: 3I + 4J + 4F + 1M (or more) = 12 (or more).

Part III—Maximum Rate of Growth
Increasing number of adult females; local population growing or emigrating; no preadult mortality.

Age in Years							
I 0–1	J 1–2	J 2–3	F 3–4	F 4–5	F 5–6	F 6–7	F 7–8
I M F M F	J M F M F	J M F M F	F	F	F	F	$F^\dagger$
			F:J + I = 1:3.0				

Average troop size: 4I + 8J + 4F + 1M (or more) = 17 (or more)

The average troop size generated by the "7-year model" would be (2I + 4J + 4F + 1 or more M) at least 11.

If, after averaging data from a number of troops, a population with the above characteristics had an F:J + I ratio of less than 1:1.5, it would be presumed it is under some stress, preventing reproductive success. A troop or population showing such a ratio over a number of years would probably show other evidence of decline (in the absence of immigration). The F:J + I ratio summarizes the status or condition of the whole subadult population, not just one small portion of that population. The F:J + I ratio contains an extended time parameter and may signal a trend in population.

Several variations are shown. Part II of Table 1 demonstrates the effect of delaying some of the mortality, i.e., permitting an extra infant to survive during age 0–1. In such a case, more subadults would be required to indicate a stable population. A third possibility is an increasing population with a stable number of adult females in a troop. This is illustrated in Part III of Table 1. Here all the infants and juveniles remain alive and half emigrate to form new troops on reaching maturity. The F:J + I ratio becomes 1:3, which is the maximum obtainable under the conditions assigned.

A second series of illustrations (Table 2) reveals the usefulness of simulation with such models. The adult female is (a) allowed 8 safe reproductive years and (b) bears one infant every other year, the "A" females reproducing one year and the "B" females the next. All the other values remain as in Part I of Table 1. This model will be known as the "11-year model." A stable population with such characteristics has an F:J + I ratio of 1:0.75 on the average. An F:I ratio of 1:0.25 is generated along with an average troop size of at least 15 (2I + 4J + 8F + 1 or more M). At the maximum rate of growth (no subadult mortality), such a popula-

TABLE 2 Models for Stable or Increasing Populations

Models for 8 years of active reproduction for females reaching maturity after 3 preadult years; 1 infant/2 years/female; 1:1 sex ratio among infants; no sex differential in preadult mortality. † indicates occurrence of a death.

Part I—11-Year Model
Number of females is constant; local population size constant; all preadult mortality assessed at birth.

	Age in Years											
I 0–1	J 1–2	J 2–3	F 3–4	F 4–5	F 5–6	F 6–7	F 7–8	F 8–9	F 9–10	F 10–11	F 11–12	
M	M	M	A	B	A	B	A	B	A	B	A†
F	F	F									
M†											
F†											

$$F{:}J + I = 1{:}0.75$$

Average troop size given eight adult females per troop: $2I + 4J + 8F + 1M$ (or more) = 15 (or more).

Part II—Maximum Rate of Growth
Increasing number of adult females; local population growing or emigrating; no preadult mortality.

M	M	M	A	B	A	B	A	B	A	B	A†
F	F	F									
M	M	M									
F	F	F									

$$F{:}J + I = 1{:}1.5$$

Average troop size: $4I + 8J + 8F + 1M$ (or more) = 21 (or more).

tion or troop would have an F:J + I ratio of 1:1.5 (Part II of Table 2).

Until computer simulation is performed, we are unable to estimate confidence intervals around single observations of troops or populations. Aberrant figures are generated when sudden drastic gains or losses occur in either portion of the F:J + I ratio, but these aberrations can be recognized by recensusing. A series of censuses will reveal when gains occur (as during immigration) and unusual drops in a particular age category. Therefore, we believe that the F:J + I ratio can be used, together with other indicators, to provide useful estimates of the status of a population.

Knowledge of howler biology, however, is spotty concerning the values that might actually be assigned to the categories outlined above. The maximum published longevity record for a captive howler is 3 years, 9 months (Napier and Napier, 1967). The current longevity record of 14 years (or more) is held by a male kept in captivity on BCI. His female consort is little more than half his age. Thus the age spans of 7 and 11 years used in the above models are perhaps within the range of typical life expectancies for captive howlers.

Carpenter (1964) estimates at least 3 years as the duration of infant and juvenile periods of life for *A. palliata*. There is no reason to assume a sex ratio at birth different from 1:1 or unequal postnatal mortality (except perhaps for males as they are about to attain sexual maturity). It seems reasonable to conclude that a conservative estimate of the F:J + I ratio for a stable population of howlers would be in the vicinity of 1:1. It could be no less than 1:0.75 and might be as high as 1:1.5. Reduction of female life span, increased length of juvenile–infant period, and postponement of mortality all serve to increase the J + I portion of the ratio necessary to maintain a stable population.

Southwick and Siddiqi (1968) present data for 41 groups of urban rhesus monkeys (*Macaca mulatta*) in India. Adult rhesus females are thought to live well beyond 11 years and reproduce each year. Nevertheless, the average proportions are 4.0 M, 9.1 F, 7.1 J and 7.1 I, yielding an F:J + I ratio of 1:1.56 and an F:I ratio of 1:0.78. Southwick and Cadigan (1972), studying habituated longtailed macaques (*Macaca fascicularis*) in Penang, Kuala Lumpur, and Cape Rachado, Malaysia, and in Singapore, report an average troop size of 24 composed as follows: 3.4 M, 8.4 F, 7.8 J, and 4.2 I. The F:J + I ratio is 1:1.43 and the F:I ratio is 1:0.50.

STUDY SITES AND PROCEDURES

The OTS class projects on the mantled howler monkey (*Alouatta palliata*) were conducted primarily on Finca Taboga, a research facility of the Costa Rican Department of Agriculture. A second field site was Finca La Pacífica, a private farm owned by Ing. Werner Hagnauer. Finca Taboga is located 11 km southwest of Cañas, Guanacaste Province, Costa Rica. Finca La Pacífica is about 7 km northwest of Cañas. Both sites are located in the tropical dry forest region of northwestern Costa Rica. On Finca La Pacífica, trees occur in windbreaks between fields, in galleries along rivers, and in larger stands on hillsides too rocky for profitable farming. Finca Taboga offers large patches of continuous forest on steep hillsides and in moist alluvial lowlands. During the dry season (January–June), the leaves fall from all but a few species of trees. Riparian galleries and swampy lowland stands are exceptions and are evergreen or only semideciduous throughout the year. The majority of the observations reported were made in the riparian gallery along the Río Higuerón at Taboga.

In any given field course, the observations were usually made on a single day over a period of 3–5 hours. The troops, sites, and number of hectares sampled were not identical from year to year. This accounts for some of the variation in total number tallied during each project and places limitations on direct comparisons between the raw figures from the censuses. Immigration from surrounding cutover regions into the study area may also have occurred since figures from one census are sometimes insufficient to account for the numbers of juveniles and females seen 3 months later or during the next year.

In each project, attention was directed toward troop size and the age and sex composition of each troop. The categories of adult male, adult female with or without infant, juvenile, and infant were generally used or could be determined from the data, which included additional categories of subadult (male, female, or unknown sex), dependent infant, independent infant, and unsexed adults.

In this paper, the definition of infant is limited to dependent infant, the Infant 1 or Infant 2 defined by Carpenter in 1934 (Carpenter, 1964). The definition of juvenile includes all nondependent subadults. However, old juveniles and young females may be interchangeably misclassified (Carpenter, 1964). We can find no reason to believe that, overall, more females

are misclassified as juveniles than juveniles misclassified as females.

The 1967 and July 1968 census totals allocate "unsexed adults" (10 and 5, respectively) according to the proportion of adults of known sex. In 1967, one troop of 28 individuals was simply counted. These individuals were divided proportionately into all four categories. "Unsexed adults and subadults" (18 animals) in the 1966 census were allocated according to the proportion of definitely determined adults and juveniles, and the adults apportioned to sexes as above. Nine "unclassified" individuals from February 1968 were divided between the known females and the juveniles. An isolated pair of juveniles in the February 1968 census were not counted as a troop nor included in the juvenile total. Lone males, encountered in several censuses, were neither counted as troops nor added to the male sum.

RESULTS AND COMPARISONS

The first portions of Table 3 present the results of nine successive censuses by OTS classes on Taboga and three censuses at La Pacífica. Taboga and La Pacífica average troop sizes are small compared to most BCI census averages except for the crash population censused by Collias and Southwick (1952). The range and standard deviation of the troop size are similar in the three areas (Table 4). The most depressed censuses (BCI, 1951, and Taboga, July 1968) show reduced range and standard deviation of troop size, though later BCI figures (1967 and 1972) are also low in these respects.

The average troop size at Taboga shows a declining trend. The correlation coefficient, r, between years elapsed (1966 = 0) and average troop size is $r = -0.61$ (the null hypothesis has a probability, P, less than 0.09, with 7 degrees of freedom, $d.f.$). The actual rate of decline in average troop size is -0.68 individuals per year. If the low average of 8.9 from July 1968 is rejected as showing severe influences not part of the general trend (see below), the correlation coefficient becomes $r = -0.76$ ($P < 0.05$, 6 $d.f.$), with an actual rate of decline of -0.75 individuals per year.

The 1932, 1933, 1935, and 1972 censuses from BCI are very similar to each other in total proportions (M:F:J:I), but differ considerably from those developed from the Taboga counts (Table 3). The 1951, 1959, and 1967 BCI censuses yield proportions that overlap Taboga values from some years. The censuses most similar in average troop size are Taboga (1966) and BCI (1972). However, the proportions of females, juveniles, and infants in these two censuses are markedly different. A chi-square test indicates that there is a very low probability that the two samples might have

TABLE 3 Population Parameters of Howler Monkey Troops at Finca Taboga and Finca La Pacífica, Guanacaste Province, Costa Rica, and Barro Colorado Island (BCI), Canal Zone, Panama

Location / Date of Study (Reference)	Average Troop Size	Number of Troops	Total	Adult Males (M)	Adult Females (F)	Juveniles (J)	Infants (I)	Total (%) M	F	J	I
Taboga											
Feb. 1966 (this study)	15.4	7	108	18	54	33	3	16	50	31	3
Feb. 1967 (this study)	13.5	10	135	23	64	38	10	17	48	28	7
Feb. 1968 (this study)	10.0	8	80	13	38	22	7	16	47	28	9
July 1968 (this study)	8.9	8	71	18	30	17	6	25	42	24	8
Feb. 1969 (this study)	13.1	17	223	52	111	37	23	23	50	17	10
Feb. 1970 (this study)	10.4	7	73	22	37	8	6	30	51	11	8
July 1970 (this study)	10.8	8	86	18	52	5	11	21	60	6	13
April 1971 (this study)	11.3	22	248	49	107	65	27	20	43	26	11
June 1971 (this study)	9.9	15	148	33	64	21	30	22	43	14	20
AVERAGES	11.5	—	—	—	—	—	—	21	48	21	10
La Pacífica											
Feb. 1967 (this study)	13.2	5	66	9	37	15	5	14	56	23	8
Feb. 1969 (this study)	10.0	6	60	19	22	10	9	32	36	17	15
Feb. 1970 (this study)	12.5	4	50	8	24	10	8	16	48	20	16
AVERAGES	11.9	—	—	—	—	—	—	20	47	20	13
BCI											
1932 (Carpenter, 1964)	17.3	23	398	63	171	92	72	16	43	23	18
1933 (Carpenter, 1964)	17.5	28	489	82	192	117	98	17	39	24	20
1935 (Carpenter, 1964)	18.2	15	273	49	105	81	38	18	38	30	14
1951 (Collias and Southwick, 1952)	7.9	30	239	36	135	32	36	15	57	13	15
1959 (Carpenter, 1964)	18.5	44	814	146	402	135	131	18	49	17	16
1967 (Chivers, 1969)	14.7	12	176	40	72	34	30	23	41	19	17
1972 (Thorington *et al.*, unpubl.)	15.2	12	182	33	66	40	43	18	36	22	24

been drawn by chance from the same population (chi-square = 23.64, $P <0.001$, 2 $d.f.$) It must be noted that the infant and juvenile classes were defined differently in the two studies. Nevertheless, if the two classes, juvenile and infant, are summed together, the composite category covers the same age range in the BCI studies as in the Taboga observations. If the summation of $J + I$ is carried out and the test is performed again on the same two groups (Taboga, 1966, and BCI, 1972), the results yield a chi-square = 5.64 with a probability of less than 0.02 with 1 degree of freedom. Comparison of the 1972 BCI data with the Taboga census next most similar in average troop size (February 1967) gives a chi-square = 4.0 ($P <.05$; 1 $d.f.$), even if the juvenile and infant categories are lumped.

The values of the ratio $F:J + I$ are presented in Table 4. Values of 1:0.75 and 1:1.5 are taken as minimal indicators of stable populations for the 11-year model and 7-year model, respectively. Even with the 11-year model, we would be more confident about the future of a population with an $F:J + I$ ratio of 1:1,

which would allow for some delay in preadult mortality. At Taboga, the ratios skirt the minimal replacement value for the 11-year model (except for 1969 and 1970) and never approach that for the 7-year model. The three La Pacífica readings suggest the possibility of a more stable situation than occurs on Taboga. This agrees with our evaluation of the relative degree and rate of environmental alteration at the two sites. In contrast to the Taboga ratios, the $F:J + I$ values from BCI vary around 1:1, except for the 1951 and 1959 values. The 1959 figures may simply indicate a downward phase in an oscillation superimposed on what appears to be a continuing upward trend in population size for howlers on BCI. Except for the depression year of 1951, less than 5 percent of the troops censused on BCI are without juveniles or infants. At Taboga, 10–50 percent of the troops counted had no infants or juveniles in six of the nine censuses. Of those five cases at Taboga in which the $F:J + I$ ratio does equal or exceed 1:0.75, females as well as juveniles and infants appear to be experiencing a reduction of numbers in February and July 1968 and possibly also in

TABLE 4 Standard Deviation and Range of Troop Size, and Indicator Ratios

Location Date of Study (Reference)	Number of Troops	Mean Troop Size ± SD	Range of Troop Sizes	F:J + I (1:—)	Number of Troops without J or I (%)	F:I (1:—)	Number of Troops without Dependent I (%)	M:F (1:—)	Number of One Male Troops (%)
Taboga									
Feb. 1966 (this study)	7	15.4 ± 12.6	3–39	0.67	0	0.06	5 (71)	3.00	2 (29)
Feb. 1967 (this study)	10	13.5 ± 7.9	5–28	0.75	1 (10)	0.16	5 (50)	2.78	3 (30)
Feb. 1968 (this study)	8	10.0 ± 6.2	2–21	0.76	1 (12)	0.18	4 (50)	2.92	4 (50)
July 1968 (this study)	8	8.9 ± 2.5	5–11	0.77	2 (25)	0.20	5 (62)	1.88	1 (12)
Feb. 1969 (this study)	17	13.1 ± NR[a]	NR[a]	0.54	NR[a]	0.21	NR[a]	2.13	NR[a]
Feb. 1970 (this study)	7	10.4 ± 5.3	5–21	0.38	1 (14)	0.16	2 (29)	1.68	1 (14)
July 1970 (this study)	8	10.8 ± 8.2	2–29	0.31	4 (50)	0.21	5 (62)	2.89	2 (25)
April 1971 (this study)	22	11.3 ± 5.6	3–26	0.86	3 (14)	0.25	5 (23)	2.18	8 (36)
June 1971 (this study)	15	9.9 ± 4.1	5–19	0.80	0	0.47	1 (7)	1.94	4 (27)
AVERAGES	—	11.5 ± 2.1	—	0.65	—	0.21	—	2.38	—
La Pacífica									
Feb 1967 (this study)	5	13.2 ± 9.1	5–27	0.54	2 (40)	0.14	3 (60)	4.11	2 (40)
Feb. 1969 (this study)	6	10.0 ± NR[a]	NR[a]	0.86	NR[a]	0.41	NR[a]	1.16	NR[a]
Feb. 1970 (this study)	4	12.5 ± 8.7	5–25	0.75	0	0.33	0	3.00	2 (50)
AVERAGES	—	11.9 ± 1.7	—	0.72	—	0.29	—	2.76	—
BCI									
1932 (Carpenter, 1964)	23	17.3 ± 7.1	4–35	0.96	1 (4)	0.42	2 (9)	2.71	4 (17)
1933 (Carpenter, 1964)	28	17.5 ± 7.0	4–29	1.12	0	0.51	3 (11)	2.34	3 (11)
1935 (Carpenter, 1964)	15	18.2 ± 7.1	6–34	1.13	0	0.36	0	2.33	2 (13)
1951 (Collias and Southwick, 1952)	30	8.0 ± 3.6	2–17	0.50	4 (13)	0.27	14 (47)	3.75	24 (80)
1959 (Carpenter, 1964)	44	18.5 ± 9.4	3–45	0.66	2 (5)	0.33	8 (18)	2.75	6 (14)
1967 (Chivers, 1969)	12	14.7 ± 2.5	11–18	0.89	0	0.42	0	1.80	0
1972 (Thorington *et al.*, unpubl.)	12	15.2 ± 4.2	8–23	1.26	0	0.65	0	2.00	0
AVERAGES	—	15.6 ± 3.7	—	0.93	—	0.42	—	2.53	—

[a]NR = data not recoverable.

June 1971. In these situations, the F:J + I ratio stays at or near the level held prior to the beginning of the generalized decline in population, but low values of the ratio follow in later years (see Table 4, *Taboga*, 1969 and 1970).

When the Taboga census figures are compared from year to year (Table 3), the low figures of the 1968 OTS observations are striking. A check with health authorities in Guanacaste Province revealed no epidemics or even minor outbreaks of yellow fever or other diseases over the 1968–69 period. Inquiries led to a veterinarian who had been assigned in 1968–69 to Liberia, Costa Rica, 48 miles northwest of Cañas. Several dead howlers that had been brought to him for autopsy showed no evidence of bacterial or viral disease, but he noted that "the gastric, stomach associated glands were swollen, puffy and inflamed." Government laboratories in San José reported that the glands were loaded with insecticides, variety not

specified. The veterinarian confirmed that 1968 and 1969 were heavy cotton years in the whole area, with large doses of insecticides used to protect the crops. The insecticides were often sprayed from airplanes, increasing the possibility of winds carrying the poisons from the fields to adjacent forested areas.

In February 1968, the first categories to decrease were the adult males, adult females, and juveniles. By July 1968, the adult females and juveniles showed a further decline from the 1967 figures. The drop in the juvenile category may have been due to the low number of infants in previous years or to insecticide mortality. Censuses subsequent to 1968 appear to indicate some resurgence, but average troop size never reached the 1967 level (Table 3). Most categories show rises in raw figures in 1969, even though dead monkeys were found at Taboga. It is probable that both an increased area of census and immigration into the census area accounted for the increased total count in

1969. Nevertheless, the proportion of juveniles continued to drop, reaching a nadir in July 1970, despite the high number of infants in 1969 both actually and proportionally. During these years, the number of females began to climb again and was accompanied by a larger infant population. With the increase in the number of infants in 1970, the number of juveniles jumped in 1971. However, the female population fell proportionally, perhaps due to the relatively high number of juveniles currently and the low number of juveniles available during the previous years. The adult females of June 1971 appeared to be very active reproductively.

If indeed the ratio F:J + I is important, a low proportion of J + I over a period of years should show up as a decrease in the female category and then again in the J + I categories because of reduced numbers of females. The Taboga figures (Tables 3 and 4) appear to illustrate such a situation.

Carpenter suggests that a low percentage of females with infants may signal a population in difficulty. Carpenter's (1964) figures show that it is usual to have from one-third to one-half or more of the adult females associated with infants. In fact, even during the census of Collias and Southwick shortly after yellow fever apparently decimated the BCI howler population, the F:I ratio dropped only slightly lower than the more usual values. Except for 1971, less than a quarter of the females at Taboga were carrying infants. The 11-year model predicts that an average of 25 percent of the adult females should be with dependent infants, and the 7-year model predicts 50 percent. Up to 70 percent of the troops censused at Taboga showed no dependent infants. In 1967 at La Pacífica, three of the five troops (60 percent) contained no dependent infants and two of these contained no juveniles. On BCI only very small troops (e.g., 1 M, 2 F) had no young (Carpenter, 1964). At Taboga troops with 4, 5, and even 9 adult females yielded counts with no infants (OTS files, San José, Costa Rica).

Carpenter (1964) suggests that a high proportion of troops with only one male and elevated numbers of adult females per adult male (low M:F ratio) might be characteristic of a recent dramatic decrease in population. These criteria derive from the Collias and Southwick census of BCI howlers in 1951. Combined with an M:F ratio of 1:3.75, 80 percent of their troops showed only one adult male (Table 4). Other BCI counts yield no more than 2.75 adult females for each adult male, with less than a quarter of the troops having only one adult male. The M:F ratio is not markedly nor consistently depressed at Taboga (Table 4), and the July 1968 value differs greatly from that of Collias and Southwick. La Pacífica censuses yield both the highest and

lowest ratios reported to this date. The troops that have only one male are more frequent at Taboga and La Pacífica than on BCI (except for the 1951 depression). Again the lowest proportion of troops with only one male was found in July 1968; but this census and the April 1971 count both yielded a troop in which no males were observed (OTS files).

Two La Pacífica troops were censused in both 1969 and 1970 (Table 5). La Pacífica II showed an F:J + I ratio changing from 1:0.80 to 1:0.33 and became a single male troop. La Pacífica III showed an opposite change in the F:J + I ratio, but the value attained just equals 1:0.75. The ratio of adult males to adult females changed from 1:2.8 to 1:3 in La Pacífica III and from 1:2.5 to 1:6 in La Pacífica II. Sixty percent or more of the adult females were without dependent young in both years.

In the dry-season censuses of 1969 and 1970, the troop counts from supposedly less-favorable areas were separated from counts made in more-favorable habitats (OTS files). The "less-favorable" habitats in 1969 were dry, deciduous, hillside forests with a minimum of riparian trees, where an average troop size was 9 compared to 15.4 in the "more-favorable" moist forest on the alluvial lowlands. By contrast, in the riparian–dry hillside forest, 6 of 23 adult females (26 percent) were with infants and F:J + I was 23:21 (1:0.91). In the moist forests only 17 of 88 females (19 percent) were with infants, and the F:J + I ratio was 88:39 (1:0.44). The M:F ratio was between 1:2 and 1:2.5 in both habitats. The La Pacífica counts from 1969, from riparian strips and strips of dry forest remaining between cultivated fields, were similar to the riparian–dry hillside data from Taboga. The poorer statistics from the "more-favorable" Taboga lowland forests are probably related to the rapidly progressing clearing in that area.

The February 1970 census considered differences between dry areas and wet areas (including riparian forests). From the Taboga area, only one dry hillside troop was studied. It yielded 4 M, 7 F, and only 1 I, suggesting harsh pressures preventing successful re-

TABLE 5 Organization for Tropical Studies Comparison of the Same Troop Over Two Years

Troop	Year	Total	M	F	J	I
La Pacífica II	1969	11	2	5	2	2
	1970	9	1	6	1	1
La Pacífica III	1969	27	5	14	4	4
	1970	25	4	12	5	4

production. The two dry La Pacífica areas yielded 5 M, 18 F, 6 J, and 5 I. The two counts from the "wet areas" of La Pacífica were, by contrast, 3 M, 6 F, 4 J, and 3 I. The La Pacífica figures were markedly better than the Taboga values and the "wet" areas better than the "dry," even at Taboga, though one of the Taboga "wet area" troops was composed of 4 M, 5 F, and no J or I (OTS files). The differences between Taboga and La Pacífica probably reflect the greater intensity of ongoing environmental disturbance at Taboga.

CONCLUSIONS

The censuses of howler monkeys at two sites in the tropical dry forests of Costa Rica present a picture of a distressed and declining population. In some respects, the howlers of Taboga and La Pacífica are even more seriously depressed than the 1951 population on BCI, which had just been decimated by yellow fever. In other characteristics, the howler population of Guanacaste Province is similar to that stressed population or intermediate between it and the subsequently censused BCI population. Of particular concern are the relatively low numbers of juveniles and infants. Theoretically, a female howler should be impregnated shortly after she becomes physiologically able to conceive (either after rearing an infant into the juvenile period or losing an infant prior to that time). Without ruling out the possibility of a birth season, we deduce that births and impregnations may occur at any time of year. This agrees with the conclusion of Carpenter (1964) from his BCI studies. However, the low numbers of infants relative to adult females at Taboga and La Pacífica suggest that fertilization or gestation are often unsuccessful or that early infancy is a period of heavy mortality. Even if an infant survives through its period of dependency, it apparently has a poor chance of reaching adulthood in most years.

It is of interest to calculate, given the suggested models, how many infants and juveniles would be expected in the various years. For instance, in 1966 the 7-year model would call for at least 27 infants as compared to the 3 observed and 54 juveniles as compared to the 33 observed. The 11-year model suggests 14 infants and 27 juveniles. These calculations allot all preadult mortality at birth. If the assessment of preadult mortality were extended in a more natural fashion, more than 81 young animals would be expected given the 7-year model. Under the 11-year model, more than 41 preadults would be expected, producing an F:J + I ratio more like that found on BCI and among macaques (Southwick and Cadigan, 1972; Southwick and Siddiqi, 1968). If the F:J + I composi-

tion of undisturbed stable howler populations is in the realm of those generated by the 7- and 11-year models, then it is clear that the Taboga population is in continuing difficulty.

Comparisons between the two Costa Rican sites and among habitats within sites present suggestive differences. The moist alluvial forests in the Taboga lowlands would appear to be very favorable habitats for howlers, while the riparian strips and windbreaks at La Pacífica appear to be only marginal. However, La Pacífica supports populations that seem to be more stable than at Taboga. The key may be that the remaining forested areas on La Pacífica are not being altered further by human intervention, while the progressive cutting of the Taboga forest causes or necessitates reduction of the population size.

Deforestation may produce many pressures on the howler population. The continual cutting progressively reduces the total resource area and probably leaves some habitable areas without monkeys, because the howlers are unable to reach these patches by tree pathways. Emigration from cutover areas may produce a transiently high population density in adjacent forest. This high density could be misleading unless one knows the history of the local forest and the numbers in the various age–sex classes of the population over the previous several years (density estimates are not available for the Costa Rican sites). Chivers (1969) suggests that there is an adjustment mechanism available to howler monkeys when their population density increases but does not yet actually exceed the capacity of their environment. He hypothesizes that under the influence of social pressures (mainly the avoidance of aggressive interactions), the number of troops becomes slightly more numerous and troop size and range diminish slightly. Yet, Chivers' hypothetical population would be increasing, which means that status indicators such as the F:J + I ratio would remain well above minimal replacement levels.

Social factors may be operative in population readjustment at Taboga, where howlers were seen fighting, and fresh wounds and scars were observed (OTS files, 1971). However, the indicators are often so depressed that additional advantages or pressures must dictate small troop size. For example, the folivorous howlers normally graze very lightly in any one tree. Given this norm, we suggest that smaller groups can forage more efficiently. With increased troop size, more trees must be visited and more time and energy spent in obtaining food. This leads to decreased feeding efficiency, which eventually may limit troop size in any environment. The limitations may come into effect earlier in an altered setting, where several troops may be forced to visit the same food sources. Where the arrangement of trees is in strips, as in the windbreaks and narrow riparian stands at La Pacífica, foraging is mostly linear. Efficiency of foraging in this restricted topography may also call for a lower average troop size than would be found in a continuously forested area permitting travel in any direction.

Part of the precarious condition of howlers in this study is almost certainly due to disturbance of the habitat by deforestation. However, even in the absence of human disruptions, it is possible that during most years the Costa Rican tropical dry forest is a marginal habitat for the howler monkey. In a series of "normally bad" years, the population total and the average troop size may decrease steadily, and the signal ratios may indicate reproductive activity slightly above or below maintenance levels. It is easy to visualize a continually decreasing average troop size with F:J + I ratios considerably more favorable than those that we have designated as minimal replacement values. In a marginal environment, a population may depend on near-maximum reproductive success in the sporadic "good" years to recoup losses. Deforestation, insecticide toxicity, and epidemics of disease sharply accentuate the marginal character of the habitat. During or subsequent to such assault, the J + I portion of the F:J + I ratio may indicate drastic reduction in population size and status.

Clearly, the population parameters can only diminish to certain levels. Below these levels, the temporal patchiness of the environment will take on the nature of a random factor, because small troops may

no longer be able to survive until a good year. The population as a whole may be unable to stabilize its numbers or prevent further decline. Perhaps the Taboga data illustrate the descending limb of a cyclic phenomenon related to marginal habitat; but if so, it is not apparent that the depth of the oscillation has yet been reached. Even very unsophisticated modeling suggests that some parameters of the Taboga population may be approaching threshold levels.

In general, to make the most meaningful estimates of the status of a particular population or species, and the conditions under which it can survive, major advances are required in theoretical and field investigations. Table 6 presents a compilation (Napier and Napier, 1967) of the current knowledge regarding age at maturity, longevity, and the duration of gestation and lactation for New World primate genera. The sparse data are almost exclusively from captive situations and include no indication of the variation that might be expected between species or between sexes. Adequate information on the life history variables distinctly depends on extended studies of local populations in which animals are individually identifiable. Scott and co-workers are beginning to collect such information by recensusing howler troops containing marked individuals at La Pacífica. Freese, working throughout the year in the tropical dry forest of northwestern Costa Rica, is developing statistics on within-year variation in troops (see pp. 4–9). However, every genus must be examined and, in many genera, individual species deserve study; for among sympatric species the varia-

TABLE 6 Gestation, Lactation, Maturity, and Longevity Data for New World Primates[a]

Genus	Gestation	Lactation	Age to Maturity	Record Longevity (yr, mo)
Callithrix	140 days	6 mo	14 mo	12 yr
Cebuella	—	3 mo	—	4 yr, 11 mo
Saguinus	140–145 days	—	—	9 yr, 10 mo
Leontopithecus	132–134 days	3 mo	—	10 yr, 4 mo
Callimico	150–170 days[b]	—	~1 1/2–2 yr [b]	4 yr, 9 mo
Cebus	ca. 180 days	—	—	40 yr
Saimiri	168–182 days	—	—	10–20 yr
Aotus	—	—	—	11 yr, 7 mo
Callicebus	—	—	—	—
Pithecia	—	—	—	13 yr, 8 mo
Chiropotes	—	—	—	15 yr
Cacajao	—	—	—	8 yr, 9 mo
Alouatta	—	18–24 mo	3 1/2–4 yr	3 yr, 9 mo
Ateles	ca. 139 days	—	—	20 yr
Lagothrix	ca. 225 days	12 mo+	4 yr	13 yr
Brachyteles	—	—	—	—

[a] SOURCE: Napier and Napier, 1967.
[b] Data from personal breeding colony (Heltne, unpublished).

tions in the life history parameters and behavioral patterns strongly relate to the exquisite division of resources in the tropical forest ecosystems.

Computer simulation can produce estimates of stable populations given any particular model. Even at such an unsophisticated level, simulation may interact in an extremely useful fashion with field checks of the status of local populations. In order to attain their true potential, however, simulations must incorporate feedback from long-term field investigations. Only with detailed statistical information on the life history parameters of undisturbed populations can computer simulation demarcate the evolutionary strategies available to primate species and the conservation strategies available to us.

Deforestation and other human assaults on tropical forests are producing dire conditions for many primate populations. It is urgent that the governments of Central and South American countries, environmentalists, conservationists, and the biomedical community and its animal suppliers express their concern in this matter in terms of increased attention to and support of the kinds of primatological studies suggested in this volume.

ACKNOWLEDGMENTS

We wish to express our gratitude to OTS field leaders Gordon Orians and Christopher C. Smith and to several dozen OTS student researchers, especially Darcy B. Kelley. We thank Richard W. Thorington, Jr. (U.S. Museum of Natural History) and Miguel A. Schön (Johns Hopkins University) for reading the manuscript and for allowing us to use unpublished data collected with the support of the Smithsonian Environmental Sciences Program. Charles Southwick, Jill Wolhandler, and Tom Sayvetz (all of Johns Hopkins University) ably criticized drafts of this paper. For the information on possible insecticide mortality, we are indebted to Dr. Román Miguel, Veterinario, Centro Agricola Veterinario Regional de Liberia, Costa Rica, currently head of the rabies project, Zoonosis Department, Ministerio Salubridad Pública, República de Costa Rica. The fieldwork was supported by National Science Foundation grants to the educational program of the Organization for Tropical Studies.

REFERENCES

Carpenter, C. R. 1964. Naturalistic behavior of nonhuman primates. The Pennsylvania State University Press, University Park. x + 454 pp.

Chivers, D. J. 1969. On the daily behavior and spacing of howling monkey groups. Folia Primatol. 10:48–102.

Collias, N. E., and C. Southwick. 1952. A field study of population density and social organization in howling monkeys. Proc. Am. Philos. Soc. 96:143–156.

Hall, E. R., and K. R. Kelson. 1959. The mammals of North America, vol. I. The Ronald Press Company, New York.

Napier, J. R., and P. H. Napier. 1967. A handbook of living primates. Academic Press, New York. xiv + 456 pp.

Smith, J. D. 1970. The systematic status of the black howler monkey, *Alouatta pigra* Lawrence. J. Mammal. 51:358–369.

Southwick, C. H., and F. C. Cadigan, Jr. 1972. Population studies of Malaysian primates. Primates 13:1–18.

Southwick, C. H., and M. R. Siddiqi. 1968. Population trends of rhesus monkeys in villages and towns of northern India, 1959–1965. J. Anim. Ecol. 37:199–204.

PRIMATE POPULATIONS IN CHIRIQUI, PANAMA

John D. Baldwin *and* Janice I. Baldwin

INTRODUCTION

The purpose of this chapter is to describe the changes in primate habitats and populations in the province of Chiriqui, Panama, over the past two decades (Figure 1). The conditions in Chiriqui are similar to other parts of Latin America that are more economically and agriculturally advanced. Although the ecological conditions and primate resources in Chiriqui may not be representative of the current conditions in poorer, undeveloped, or remote areas, they do represent one facet of the current plight of primate populations in the New World where the government has endorsed agricultural expansion rather than the protection of natural resources. The early parts of the chapter present data from primate surveys in several areas of Chiriqui. The latter sections deal with a single forest where we conducted an intensive study on primate ecology and behavior between December 19, 1970, and February 25, 1971.

CHIRIQUI

In August 1968 and December 1970, we surveyed 71 forested areas in Chiriqui, looking for primate groups (Figure 2). We obtained valuable overview information on primate locations from several hacienda owners and from Mr. Sigfrido Esquivel, who had once exported animals from Chiriqui. When working in the field, location of animals was simplified with the help of the local people, who would indicate areas where animals could be found. In general, their reports were accurate. When groups were not located as reported, it

was probably due to the animals having moved into forests or fields adjoining the area that was surveyed. Observations on the forests, troop size, and behavior were made with the naked eye and binoculars, then recorded on tape recorders or file cards, and later transcribed into field notebooks. The field procedures used are described in greater detail elsewhere (Baldwin and Baldwin, 1971, 1972, 1973).

The main purpose of these observations was to locate a good site for conducting an intensive 10-week study on New World primates in a natural forest in Central America. The 10-week study was to be focused on *Saimiri sciureus oerstedii* in order to compare their ecology and behavior with *Saimiri* that had previously been studied in the llanos of Colombia (Thorington, 1967, 1968) and in a seminatural environment in Florida (DuMond, 1968; Baldwin, 1968, 1969, 1971). While searching for forests containing *Saimiri*, we also encountered troops of *Alouatta palliata* (howler monkeys) and *Cebus capucinus* (capuchin monkeys). We found no evidence in the field that *Saimiri* lived in the mountainous northern part or in the eastern half of the province. Local people reported that *Saimiri* were to be found only in the southwest,* as confirmed by us, and reported that *Alouatta* and *Cebus* were more broadly distributed than our *Saimiri*-oriented investigations indicated. We never located or heard reference to other primate species in the areas studied, although W. C. O. Hill

*Mr. Esquivel, the man who exported animals from Chiriqui in the 1950's, said that once *Saimiri* had ranged as far east as Remedios in eastern Chiriqui. In 1968 we found no indication that *Saimiri* still ranged that far.

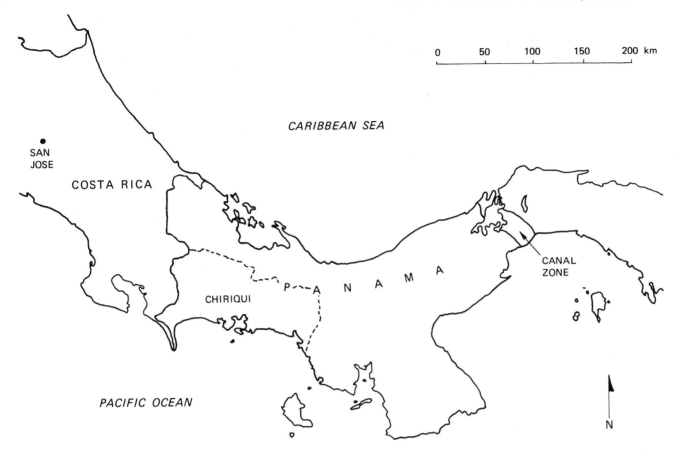

FIGURE 1 The location of Chiriqui in southwestern Panama.

(1960, 1962) indicated that *Aotus trivigatus* (night monkeys) and *Ateles geoffroyi* (spider monkeys) were once found in the province. Areas in which each species of primate was observed are indicated in Figures 3, 4, and 5.

In southwestern Chiriqui it was possible to locate primates only in limited areas. Judging by the reports of many local people, the primate populations had diminished significantly during the two decades before 1968 through (1) clearing and fumigating the land for agriculture, (2) trapping for export, and (3) hunting by people. The *Federal Register* has placed *Saimiri sciureus oerstedii* on the endangered species list (Russell, 1970), but in the areas of Chiriqui we sampled *Cebus* were more endangered.

EFFECTS OF AGRICULTURE

The clearing and fumigating of land for agricultural use and pest control has been the major cause of the destruction of primates in Chiriqui. Chiriqui is one of the largest cattle producing and finishing provinces in Panama. It also is a major producer of rice, corn,

cacao, coconuts, sugarcane, and bananas. Chiriqui contains large expanses of gently rolling and flat land with relatively high soil quality compared with other parts of Panama. In the 1920's a short railroad system was opened that made it possible to transport agricultural products to Puerto Armuelles for export by ship. In the 1950's Chiriqui became connected to the remainder of western Panama and the Canal Zone by an all-weather road. A system of sand, dirt, and some asphalt roads links various parts of Chiriqui and makes its back country more accessible to vehicles than is the case in most other agricultural provinces of Panama. Specific government policies have been inaugurated to facilitate and accelerate the economic development of the area and to encourage people to develop the land rather than to move to the cities.

Between 1950 and 1960 the population of Chiriqui increased from 138,136 to 188,350, and in the following 10 years it increased to 236,256 (Censos nacionales, 1960, 1970). This is a population growth rate of 36 and 25 percent increase per decade. As a consequence of these and other factors, there has been a rapid economic development and a strong impetus to cut

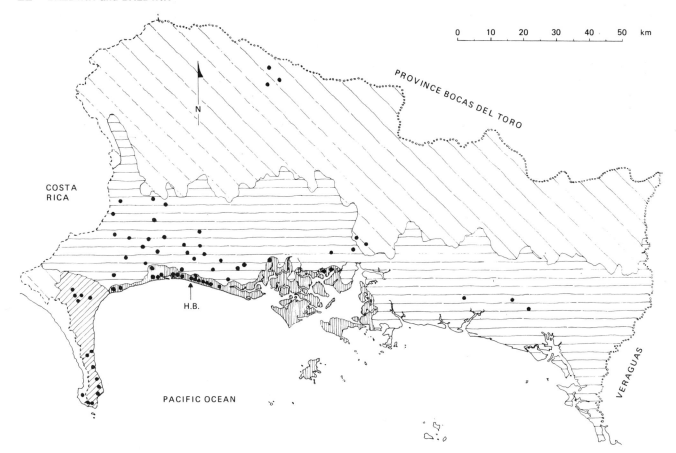

FIGURE 2 The province of Chiriqui in southwestern Panama. The forests that were investigated are indicated by dots. The four habitat zones designated by cross-hatching, from the least line density to the most, are areas over 200-m elevation, inland forests under 200 m, Burica Peninsula, and marshy coastal regions. The 10-week study conducted at Hacienda Barqueta is marked H.B.

down the forests that once covered most of the province in order to create more land for farms and ranches (Legters *et al.*, 1962). In 1950, Chiriqui was 60.7 percent forested. By 1960, the forests were reduced to 50.4 percent (Censos nacionales, 1960). Doubtlessly many thousands of primates died as their habitat was destroyed. Several people reported that pesticides and other poisons were responsible for killing large numbers of primates and other small vertebrates. Because pesticides were used on both cleared land and many forested areas, primate populations in otherwise unmolested forests were often affected.

EXPORTATION

Mr. Sigfrido Esquivel of David, Chiriqui, who is very familiar with primate resources in the province, provided us with the following overviews. Between 1952 and 1960 he operated a small business collecting and exporting primates and other exotic animals. In the early 1950's *Saimiri, Cebus,* and *Alouatta* had been

common in various areas of Chiriqui. By 1960, however, many forests had been cleared for agricultural use and many of the remaining forests had been fumigated or trapped until only refugee primate populations remained. The primates had become so scarce in the accessible parts of the province that they were no longer profitable to export. According to Mr. Esquivel, bands of monkeys still existed in numerous scattered areas; but larger, more natural troops existed only in relatively remote, inaccessible areas. Our observations in 1968 and 1970 confirmed that monkeys still exist only in scattered and/or remote areas.

Mr. Esquivel reports that he was the only exporter in Chiriqui and that between 1952 and 1960 he exported 500 *Saimiri*, 200 *Alouatta*, and 600 *Cebus* to the Canal Zone. A larger number of each species may have been trapped, however, since there are usually appreciable attrition rates before the trapped animals reach the exporters. From the Canal Zone the animals could have been exported to much farther destinations. Judging by personal observations in Panama in comparison with observations in Leticia, Colombia, and

Iquitos, Peru, it seems likely that the collecting, caging, holding, and transportation procedures used in Chiriqui and the Canal Zone between 1952 and 1960 were probably inadequate to guarantee the health of the animals. Given these conditions, it is probable that significant proportions of the animals could have died during the first months of captivity and transportation.

The primates of Chiriqui were probably not utilized in scientific research. *Cebus* and *Alouatta* have never been used extensively in laboratory research. During the 1950's, when the major exportations were made from Chiriqui, *Saimiri* was not yet the popular laboratory animal it became during the 1960's. Thus, the 1,300 monkeys that Mr. Esquivel exported from Chiriqui during the 1950's probably went to zoos and the pet trade rather than to laboratories. As Southwick *et al.*

(1970) noted, it is important for the scientific community to be aware of the various causes behind the nearly global destruction of primate resources:

We believe there is a danger of undue emotionalism about primate conservation before adequate field data are available. It is likely that biomedical research will receive the brunt of blame for many problems. When shortages of primates occur, the most convenient and visible scape goat is the research laboratory. (p. 1053)

For the research community, the practical problem right now is to attach the blame for attrition of primate populations where it belongs; . . . (p. 1054)

HUNTING

Hunting by the people is the third factor that might contribute to the destruction of primate populations in

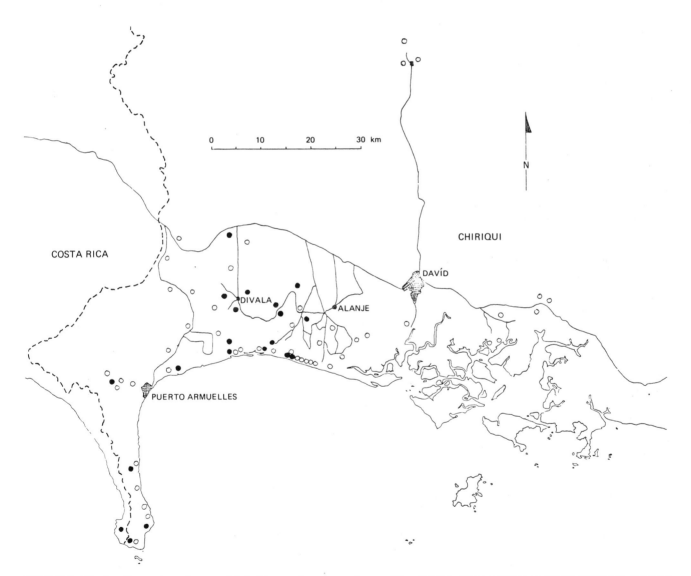

FIGURE 3 The locations where troops of *Saimiri* were found are shown with solid dots. The locations where no *Saimiri* were found are indicated by open circles. Lines represent the roads used.

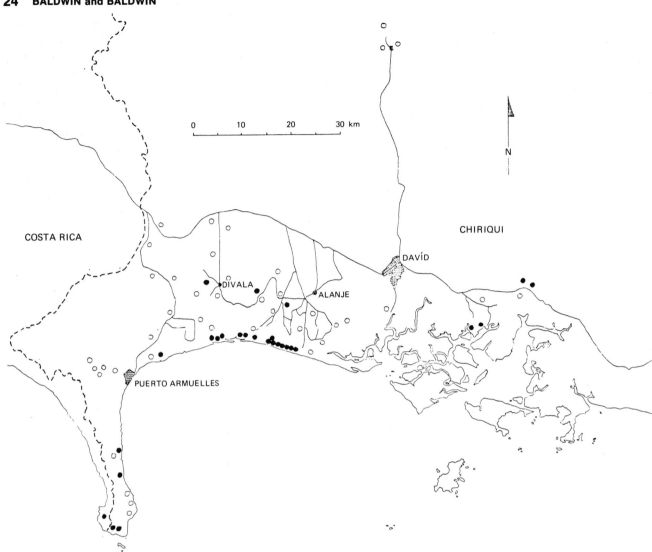

FIGURE 4 The locations where troops of *Alouatta* were found are shown with solid dots. The locations where no *Alouatta* were found are indicated by open circles.

Chiriqui. There were no indications that primates in Chiriqui were commonly hunted for food in the recent past.* There were reports that some city dwellers occasionally killed a few monkeys for "sport." The local people were generally aware of nearby bands of monkeys and had a mild positive or joking affection for them. Some people claimed that the monkeys were nuisances, and a few offered to assist us if we wanted any killed.

Of the three species, *Cebus* were generally considered to be the worst pests and were most frequently spoken of as "bad." *Cebus* stole food and crops, as has also been reported by Oppenheimer (1969). *Saimiri* occasionally stole food; but this was less common, and

*Although primates are not taken for food in Chiriqui, in parts of Amazonia and the llanos of Colombia *Cebus* and *Alouatta* are among the primates hunted for food by various indigenous peoples.

the *Saimiri* were not considered to be as destructive as the *Cebus*. It is doubtful that local hunting was responsible for much of the recent decline in the primate populations in southwestern Chiriqui, but hunting has probably affected *Cebus* the most.

The destruction of large expanses of forest habitat for agricultural expansion and the use of pesticides have been the main factors in the reduction of primate populations in Chiriqui. Exportation of primates for nonscientific use has been a minor factor. Hunting by local people is an additional minor factor that is difficult to evaluate.

THE FORESTS

Most of the three types of areas surveyed in this study were in the southwestern part of the province: the inland lowlands, the Burica Peninsula, and the marshy coastline (Figure 2).

The inland lowland areas (below 200 m or 650 ft) were the best suited for agriculture and were the most heavily exploited. The forests that remained in 1968 ranged between 0.2–40 ha (0.5–100 acres) and averaged 2–8 ha (5–20 acres) in size. Many forests were isolated by pastures and cultivated fields with few forests found near cities, towns, and the larger roads. The largest and most continuous tracts of forest were located along streams, rivers, and in less-accessible areas. Large haciendas tended to cut down sizeable sections of forest but leave untouched their more remote or less-valuable areas. Small landholders had smaller, scattered fields and often 0.2–4-ha (0.5–10 acres) forests remained between their cultivated areas.

The larger forests in undeveloped areas of large haciendas were the least-molested lowland forests observed. The small forests scattered through other areas tended to consist of much second growth, scrub, and thickets that were penetrated by numerous trails or kept clear by the foraging of domestic cattle, swine, and other animals. Riverine forests or lines of trees that extended along streams, roads, and fences often served as connecting routes between small forests, and monkeys were observed to use these as routes to move from forest to forest. Numerous forests, however, were completely isolated from other forest areas. Only a small proportion of these 0.2–4-ha (0.5–10 acres) forest "islands" contained troops of monkeys. Many wire fences used posts made of freshly cut *Spondias* saplings that took root and made living fence lines,

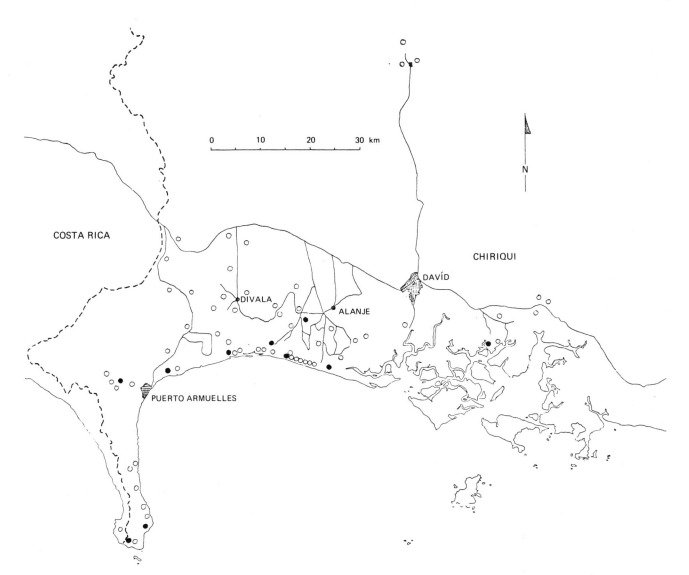

FIGURE 5 The locations where troops of *Cebus* were found are shown with solid dots. The locations where no *Cebus* were found are indicated by open circles.

which monkeys used for crossing between forests. Monkeys were also observed to cross the ground when forests were 5–20 m (16–65 ft) apart.

The Burica Peninsula is not as flat as the remainder of southern Chiriqui. A 150–200-m (500–650 ft) high mountain ridge separates the Costa Rican side from the Panamanian side, and much of the terrain is steep and cut by gullies and streams. There are flat areas near the coastline and the southern tip of the peninsula. There were no roads on the southern 24 km (15 miles) of the peninsula; but small boats could land along much of the coast, and foot and horse travel into Puerto Armuelles was common along the beach. Although there were a few large haciendas on the peninsula, there were many small farms. Considering the ruggedness of the terrain, a surprising number of forests on the peninsula had been cut down. Flat areas had been opened for crops and pastures, and steep slopes had been cleared for cattle. The Costa Rican side of the Burica Peninsula was reported to be less developed than the Panamanian side. Nevertheless, on the Panamanian side there were numerous forest areas 0.4–16 ha (1–40 acres) that contained one, two, or all three species of primates.

A distinct type of ecology exists on the islands and part of the coastline (Figure 1). Extending parallel to the coastline and up to 3.2 km (2 miles) inland were expanses of hundreds or thousands of acres broken only occasionally by streams or clearings, and penetrated by few trails. In these areas, marshes, thickets, scrub palms, trees, and mangrove swamps were common in the lower, inundated areas. Medium tall forests (12–28-m trees) (40–90 ft) could be found in higher areas, such as along old dune lines located 0.2–1.5 km (0.1–1 mile) inland from the present shoreline dunes. The poor quality of the sandy soil, marshiness, and difficulty of landing boats made much of the marshy coastline relatively inaccessible and undesirable to humans. Therefore, we found these marshy coastal areas had the largest continuous undisturbed forests. We did not study on the islands, which are mostly mangrove and marshy, since they were described as places difficult to live in or work. However, all three species of primates were reported to inhabit the islands. The marshy coastline on the mainland was more easily studied and also contained relatively high concentrations of all three species of primates. This resulted in our selection of a coastal forest for the 10-week field study on primate ecology and behavior.

THE PRIMATES

In inland lowland forests, *Cebus* traveled in small bands of two–five animals that were very wary of man

and fled in silence when approached. The wary behavior could be a consequence of their being harassed, trapped, or hunted more frequently than in larger forests. Only on the Burica Peninsula and in the coastal forests were there areas where *Cebus* traveled in groups of 20 or more; in these troops a few animals made bold displays at humans while their troops meandered nearby or drifted away.

Likewise, troops of *Alouatta* were small and scattered in the inland lowland forests compared with the Burica Peninsula and the coastal forests. On the peninsula and especially in the marshy coastal areas, *Alouatta* could be found in troops of 10–30 animals. One could often see or hear several troops at once from an observation point near the edge of a forest.

In disrupted areas *Saimiri* lived in larger troops than the other two species. In the inland lowlands, troops of 10–20 *Saimiri* could be found in forests as small as 0.8–2 ha (2–5 acres). The *Saimiri* ventured out into cornfields, banana plantations, and other crop and fallow lands in search of food or to make crossings between forests. On the Burica Peninsula and along the marshy coastlines, the *Saimiri* troops were more common and traveled in larger groups of 15–30 animals with the largest troop sizes occurring in the larger forests. If the present rate of agricultural expansion on the Burica Peninsula and in the marshy coastal forests continues, the primate populations and troop sizes will undoubtedly be diminished; and this situation will resemble the conditions in the inland lowlands.

THE 1970–1971 STUDY FOREST

The forest at Hacienda Barqueta was selected to be the locus of an intensive 10-week study (December 19, 1970, to February 25, 1971) on primate ecology and behavior, because it contained *Alouatta, Saimiri,* and *Cebus* and because it was remote from and relatively unmolested by people. The location of the study site is shown with an arrow marked H.B. in Figure 2. The study site consisted of a 20-ha (50 acres) subset of a continuous undisturbed forest of over 400 ha (1,000 acres) that extended 11 km (7 miles) parallel to the Pacific coast and varied from 0.4–1.6 km (0.25–1 mile) in width. The forest was bordered on the north by an estuary to the Escarrea River and on the south by dunes and grasslands that were flooded in wet season (Figure 6). In wet season, part of the forest floor was covered with water, but approximately 85 percent was elevated above the wet season high-water level. Surrounded by water most of the year, the study site was effectively isolated from human molestation, and species of animals rare elsewhere in southern Chiriqui were present in the forest. During the early weeks of

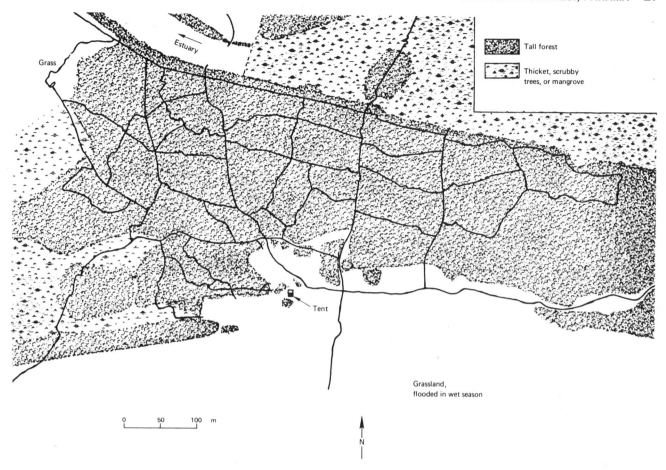

FIGURE 6 The study site at Hacienda Barqueta, showing the trail system.

the study, 8 km (5 miles) of trails were opened through the dense forest in order to follow the primates throughout the whole study area: 1 troop of *Cebus capucinus*, 2 troops of *Saimiri sciureus oerstedii*, and 11 troops of *Alouatta palliata*. The number of animals of each species totaled 27 or 30 *Cebus*, 50 *Saimiri*, and approximately 210 *Alouatta* (Table 1). This represents a very high population density of primates compared with most primate populations reported in the literature.* The compositions of *Saimiri* and *Alouatta* troops are presented in Table 2.

The remainder of the 400-ha forest could not be easily surveyed, since we could not afford the time or effort to open a trail system like the one in the study forest. However, we found similar densities of primates along the forest edge throughout the area and therefore believe that the population density found in the 20-ha study area was similar to that throughout hundreds, if not thousands, of acres of the marshy coastal forest.

*The methodology used in studying *Saimiri* is presented in Baldwin and Baldwin, 1972; the methodology for *Alouatta* is in Baldwin and Baldwin, 1973.

TABLE 1 The Population Size and Home Range Data for the Better-Studied Troops That Used the Study Site at Hacienda Barqueta

Species and Troop	No. of Animals	Range Size		Exclusive Range	
		Hectares	Acres	Hectares	Percent
Cebus					
One troop	27–30	32–40	80–100	—	—
Saimiri					
Troop A	23	17.5	43.5	1.8	10.3
Troop B	27	24–40	60–100	—	—
Alouatta[a]					
Planta	20	6.3	15.7	0.0	0.0
Boca	18	6.6	16.5	0.0	0.0
27-4-2	27	6.9	17.2	0.0	0.0
Pata	16	3.2	8.0	0.3	9.4
Saco	16	3.2	8.0	0.28	8.8
7-2-0	7	—	—	—	—
Cola	28	3.5	8.7	1.2	34.3
L.D.	19	4.2	10.5	—	—
Solitaries	6				

[a] Three additional troops of *Alouatta* used the 20-ha study site, but we were unable to collect sufficient data on them to obtain accurate troop counts.

TABLE 2 The Troop Compositions of the *Saimiri* and *Alouatta* Troops That Were Well Studied

Species and Troop	Troop Size	Adult Males	Subadult Males[a]	Adult Females[a]	Juve-niles[a]	In-fants	Percentage Ratios	Adult Females with Infants (%)
Saimiri								
Troop A	23	3	2	6	7	5	13:09:26:30:22	83.3
Troop B	27	2	3*	7*	9*	6	07:11:26:33:22	85.7
Average							10:10:26:32:22	84.7
Alouatta								
Planta	20	6	—	7*	4*	3	30:35:20:15	42.9
Boca	18	2	—	8*	5*	3	11:44:28:17	37.5
27-4-2	27	4	—	12**	6**	5	15:44:22:19	41.6
Pata	16	3	—	7	2	4	19:44:12:25	57.2
Saco	16	4	—	7*	3*	2	25:44:19:12	28.6
7-2-0	7	2	—	3*	2*	0	29:43:29:00	00.0
Cola	28	6	—	13**	3**	6	21:46:11:21	46.1
L.D.	19	4	—	7*	5*	3	21:37:26:16	42.9
Solitaries		1	1		4			
Average Troop							21:42:20:17	40.6

[a] Since we had difficulty in discriminating reliably between older juveniles and young adult females without infants, the data presented are best estimates. One asterisk indicates an estimate of ±1 accuracy. Two asterisks indicates an estimate with ±2 accuracy.

At the study site, the *Saimiri* and *Alouatta* habituated to our presence in the first 2 weeks of work; thereafter, continuous observations could be made from 7–15 m (23–50 ft) without disturbing the animals. The rapidity of this habituation process supports the impression that these two species had not been molested by man in the recent past. The *Cebus*, however, did not habituate well; even at the end of the study they became wary whenever humans approached within ranges of 15–30 m (50–100 ft). Adult *Cebus* occasionally threatened us, shook and dropped branches, and gave the gyrrah vocalization described by Oppenheimer (1968). Some local people said that two *Cebus* had been trapped nearby some years ago.

REFUGEE POPULATIONS

From the divergent stories of numerous local people it was difficult to piece together a consistent picture of the history of the area around the study forest at Barqueta. Apparently, slash and burn techniques had been used to open croplands north of the marshes for many decades. More recently, tractors and bulldozers speeded the process. However, as late as 1960 continuous or adjoining forests of up to 40 ha (100 acres) were still common throughout the area. As these forests were further destroyed in the 1960's, probably many primates became trapped in isolated small forests and others migrated into the uncut forests. Even during the study, one to four men worked several days a week with machetes and fires clearing sections

of forests near the study site. In some cases, the monkeys were trapped in isolated stands of trees where they might have starved to death. During the study, however, all three species were observed to cross between forests on the ground. Among the *Alouatta*, injured and isolated animals made terrestrial crossings up to 75 m (250 ft) through tall and short grass between forests. Healthy *Alouatta* belonging to integrated troops rarely made ground crossings longer than 5 m (16 ft). *Cebus* and *Saimiri* were not observed to make ground crossings longer than 20 m (65 ft) and 12 m (39 ft) respectively.

The high degree of separation of most of the forests in southern Chiriqui in 1970 would suggest that primate populations living more than 1.5–2.5 km (1–1.5 miles) inland from the marshy coastal forests probably could not have reached the Barqueta study forest unless they lived along the Escarrea River, which parallels a forest for several miles from the coast. There were, however, many areas adjacent to the Barqueta study forest that had once been forested, suggesting the animals retreated into the remaining forests. Every 1 km² of remaining forest in 1970 could have received monkeys from 1–3 km² of adjacent land where forests were destroyed over the past decades.

We suspect that over the years there has been a slow and gradual tendency for refugee monkeys to drift into the protected coastline forests and to create a slow rise in the population density there. The *Alouatta*, whose diet was leaves, fruits, and flowers, had adequate food supplies in the coastal forests. They apparently adapted to living in small home ranges averaging 4.8 ha (12.1 acres) with an overlap of between 65.7 and 100 percent, averaging 94 percent overlap. The population density at Barqueta was 21 times as high as the population density on Barro Colorado Island in 1932 and 12 times that of 1967 (Carpenter, 1934; Chivers, 1969). Yet, the *Alouatta* at Barqueta showed no signs of increased aggression or pathological behavior in comparison with the *Alouatta* on Barro Colorado Island. Fertility was normal: 41 percent of the adult females at Barqueta had infants compared with 37 percent on Barro Colorado Island in 1932, 49.5 percent in 1933, 26.7 percent in 1951, and 41.7 percent in 1967 (Carpenter, 1934; Collias and Southwick, 1952; Chivers, 1969). The *Alouatta* had sufficient food sources, even in dry season, when several of their food trees lost their leaves. There were few signs of botfly infestations, and all the animals (except two injured solitary juveniles) appeared to be healthy and strong.

Two troops of *Saimiri* shared the study forests. The diet of *Saimiri* consisted of insects and certain fruits. In contrast with *Alouatta*, the *Saimiri* had a very limited food supply at Barqueta, at least during the 3 months of the study. Animals of all age–sex classes in both troops spent 95 percent of each hour of the 14-hour waking day engaged in foraging and traveling between foraging locations. The *Saimiri* took no midday rest periods in the heat of the day, as they do in other environments (Baldwin, 1967; Thorington, 1968); they seldom rested more than 60 seconds at a time; and they were never observed to engage in a single bout of social play. Insect foods were rarely found, and the monkeys ate larger quantities of plant food than has been reported for the species in other environments (Fooden, 1964; Thorington, 1967, 1968). The principal food sources were the fruits of the cactuslike plant *Aechmea* and the palm *Scheelea* cf. *zonensis,* which grew in dense clusters and took 15–120 seconds to locate and/or pick. Because ripe fruits were concentrated on single plants, the whole troop often came together within a 180-m² (215 yd²) area. This contrasts markedly with the troop observed by Thorington (1967, 1968) on the llanos of Colombia, where *Saimiri* fed on small, dispersed canopy insects, and the troop tended to split up into subgroups that foraged in different directions at different rates.

As refuge groups of *Saimiri* have moved into the remaining Barqueta forests, probably starvation and disease have kept their numbers in balance. If the population changes have been gradual rather than erratic, the population density of 13 *Saimiri* per 10 ha (25 acres) may represent the maximum population density for the Barqueta ecology. The hacienda owners at Barqueta said that before the insecticide spray-

ing, *Saimiri* were much more abundant throughout the forests. The owners reported in 1971 that it had been approximately 10 years since their forests were last sprayed.

It was difficult to assess the population density of the *Cebus*. Only one troop was observed at the study site, and its home range was larger than the study area. Whether it had territorial conflicts with adjacent troops could not be determined. Oppenheimer (1968) found that *Cebus* on Barro Colorado Island traveled in groups of 15 and defended territories of approximately 85.2 ha (213 acres). At Barqueta the home range of the one troop of 27 or 30 *Cebus* was probably around 32–40 ha (80–100 acres). The larger troop size and smaller home range size may reflect crowding due to an influx of refugee populations. The *Cebus* did not appear to be as hungry as the *Saimiri,* though they did spend 50–70 percent of their day foraging and traveling. No abnormal or unusual behaviors were noted. However, a *Cebus* was observed to attack a 1.7-m (5.6 ft) iguana, wrestle with it for 30 seconds, and break off 30–40 cm (12–16 in.) of its tail. The iguana fell 12–15 m (40–50 ft) to the ground and scurried away, while the *Cebus* ate the meat off the tail fragment. This is the largest animal known to be attacked by *Cebus.*

Table 3 shows the arboreal mammalian biomass in the study forest. At Barqueta, where the population density was relatively high compared with most primate ecologies, *Alouatta* comprised 84 percent of the arboreal mammalian biomass. Eisenberg *et al.* (1972, p. 869) point out: "Not surprisingly the arboreal folivores are the most numerous of larger forest mammals, sometimes accounting for 30–40 percent of the arboreal mammalian biomass." The higher percentage at Barqueta probably reflects the high population density.

LAND CLEARING

In early 1972 a portion of the 400-ha forest surrounding and including the study site was bulldozed to the ground. The owners of the hacienda, Mr. and Mrs. Julio Arauz, wrote that they were under pressure from the local government to make "profitable" use of their forested lands or have the lands taken over under the government land reform laws. Mr. C. Neal McKinney, Administrative Officer of the Smithsonian Tropical Research Institute in the Canal Zone, has confirmed that "implementation of an old land reform law has been speeded up by the present military government" (1972, personal communication). As a consequence, the hacienda owners cut most of the tall forests on the higher ground and turned the land to cattle-grazing pasture. The Arauzes reported that some monkeys

TABLE 3   The Biomass of the Known Arboreal Mammals at the Study Site

Species	Kg/Ha	Total (%)
Alouatta (howlers)	44.60	84.0
Cebus (capuchins)	1.83	3.4
Saimiri (squirrel monkeys)	0.69	1.3
Bradypus (sloths)	0.53	1.0
Microsciurus (squirrels)	0.17	0.3
Nasua[a] (coatis)	5.25	10.0
TOTAL	53.07	100.0

[a] *Nasua* traveled, foraged, and rested arboreally around 25 percent of the daylight hours. The percent of their food intake that came from arboreal sources is unknown.

were killed during the bulldozing. Others, however, retreated into adjacent low marshy areas containing mangrove and scrub forests, where many were dying of starvation. During our field observations, it was clear that the *Alouatta* and *Saimiri* spent very little time in the mangrove and marshy scrub forests. The *Cebus*, however, spent at least 50 percent of their time in and around the mangrove areas. Although the *Cebus* may be able to adapt to the new circumstances, it can be assumed that the other two species will be more heavily affected by the destruction of their preferred habitat.

SUMMARY

In August 1968 and December 1970, the authors surveyed primate populations in 71 forested areas in the province of Chiriqui in southwestern Panama. Troops of squirrel monkeys (*Saimiri sciureus oerstedii*) were found in 20 areas, troops of howling monkeys (*Alouatta palliata*) were found in 27 areas, and troops of capuchin monkeys (*Cebus capucinus*) were found in 10 areas. The inland lowlands below 200-m altitude have been developed extensively for agricultural use, and the few primate troops that remained in these areas were small, scattered, and often afraid of man. On the Burica Peninsula there has been less agricultural development, and larger troops were more abundant. The marshy coastal areas that extend along part of Chiriqui's coast have been least affected by the rapid agricultural development in the province. These areas contained the largest primate populations.

Between December 19, 1970, and February 25, 1971, the authors conducted a 10-week study on the ecology and behavior of the three species of primates in an extensive coastal forest. The population density of *Alouatta* was 104 animals per 10 ha (25 acres), which was 21 times greater than Carpenter observed on

Barro Colorado Island in 1932 and 12 times greater than Chivers observed there in 1967 (Carpenter, 1934; Chivers, 1969). The troops' home ranges overlapped extensively, leaving an average of only 6 percent of one troop's home range for its exclusive use. There were no signs that the *Alouatta* were (1) exhausting their food supply of leaves, fruits, and flowers; (2) experiencing decreased fertility; or (3) suffering pathological effects from the high population density. The *Saimiri,* on the other hand, were living in starvation conditions at the population density of 13 animals per 10 ha (25 acres). The insects and fruits that normally comprise the *Saimiri's* diet were very scarce at Barqueta, and the *Saimiri* spent 95 percent of every waking hour engaged in foraging and traveling between foraging areas. This restricted the frequency of social interactions among them compared with other natural and seminatural environments. The *Cebus* did not habituate well to humans and could not be studied as closely as the other two species. One troop of 27 or 30 *Cebus* used about 32–40 ha (80–100 acres), creating a population density four times greater than that reported for the *Cebus* on Barro Colorado Island (Oppenheimer, 1968). Data presented suggest that the high population densities at the coastal study site may have been a consequence of the influx of refugee populations from once-forested areas to the north, as those areas opened for agriculture over the past 20 years.

In 1972 the study forest was bulldozed to the ground in order to open more land for cattle pasture. Panamanian land reform laws are encouraging the destruction rather than the conservation of many forested areas in order to encourage agricultural development and economic growth. Primate populations have been reduced significantly as their habitats have been destroyed and as pesticides have been introduced into agricultural usage. If the present rate of economic development continues and no attempt is made to protect the wildlife, the primate populations in southwestern Chiriqui will be endangered in the near future. A national park or wildlife refuge is needed to preserve some portion of the flora and fauna in these areas.

ACKNOWLEDGMENTS

We would like to thank the following persons who aided our work: Mr. and Mrs. Julio Arauz, the owners of Hacienda Barqueta, who generously permitted us to use their forests, open trails, and conduct the 10-week study without interference; Mr. Sigfrido Esquivel, who provided information on primate conditions in Chiriqui; Dr. Richard Thorington, Jr., who provided the original information on the location of *Saimiri* in Costa Rica and Panama; and Dr. Martin Moynihan, who arranged for our trip to the Burica Peninsula.

REFERENCES

Baldwin, J. D. 1967. A study of the social behavior of a semifree-ranging colony of squirrel monkeys (*Saimiri sciureus*). Unpublished doctoral thesis. Johns Hopkins University.

Baldwin, J. D. 1968. The social behavior of adult male squirrel monkeys (*Saimiri sciureus*) in a seminatural environment. Folia Primatol. 9:281–314.

Baldwin, J. D. 1969. The ontogeny of social behavior of squirrel monkeys (*Saimiri sciureus*) in a seminatural environment. Folia Primatol. 11:35–79.

Baldwin, J. D. 1971. The social organization of a semifree-ranging troop of squirrel monkeys (*Saimiri sciureus*). Folia Primatol. 14:23–50.

Baldwin, J. D., and J. I. Baldwin. 1971. Squirrel monkeys (*Saimiri*) in natural habitats in Panama, Colombia, Brazil and Peru. Primates 12:45–61.

Baldwin, J. D., and J. I. Baldwin. 1972. The ecology and behavior of squirrel monkeys (*Saimiri oerstedi*) in a natural forest in western Panama. Folia Primatol. 18:161–184.

Baldwin, J. D., and J. I. Baldwin. 1973. Interactions between adult female and infant howling monkeys (*Alouatta palliata*). Folia Primatol. 20:27–71.

Carpenter, C. R. 1934. A field study of the behavior and social relations of the howling monkeys (*Alouatta palliata*). Comp. Psychol. Monogr. 10:1–168; reprinted *in* C. R. Carpenter. 1964. Naturalistic behavior of nonhuman primates, pp. 3–92. Pennsylvania State Univ. Press, University Park.

Censos nacionales de 1960 de Panama. Panama: Dirección de Estadistica y Censo.

Censos nacionales de 1970 de Panama. Panama: Dirección de Estadistica y Censo.

Chivers, D. J. 1969. On the daily behaviour and spacing of howling monkey groups. Folia Primatol. 10:48–102.

Collias, N. E., and C. H. Southwick. 1952. A field study of population density and social organization in howling monkeys. Proc. Am. Philos. Soc. 96:143–156.

DuMond, F. V. 1968. The squirrel monkey in a seminatural environment. Pages 87–145 *in* L. A. Rosenblum and R. W. Cooper, eds. The squirrel monkey. Academic Press, New York.

Eisenberg, J. F., N. Muckenhirn, and R. Rudran. 1972. The relation between ecology and social structure in primates. Science 176:863–874.

Fooden, J. 1964. Stomach contents and gastrointestinal proportions in wild-shot Guianan monkeys. Am. J. Phys. Anthropol. 22:227–231.

Hill, W. C. O. 1960. *Primates,* vol. IV, part A. Edinburgh Univ. Press, Edinburgh.

Hill, W. C. O. 1962. *Primates,* vol. V, part B. Edinburgh Univ. Press, Edinburgh.

Legters, L. H., W. Blanchard, E. E. Erickson, B. C. Maday, N. S. Popkin, and S. Teleki. 1962. Special warfare area handbook for Panama. U.S. Government Printing Office, Washington, D.C.

Oppenheimer, J. R. 1968. Behavior and ecology of the white-faced monkey, *Cebus capucinus,* on Barro Colorado Island, C.Z. University Microfilms, Ann Arbor, Michigan.

Russell, F. J. 1970. Conservation of endangered species and other fish and wildlife. Federal Register 35:8491–8498.

Southwick, C. H., M. R. Siddiqi, and M. F. Siddiqi. 1970. Primate populations and biomedical research. Science 170:1051–1054.

Thorington, R. W., Jr. 1967. Feeding and activity of *Cebus* and *Saimiri* in a Colombian forest. Pages 180–184 *in* D. Starck, R. Schneider, and H. J. Kuhn, eds. Neue ergebnisse der primatologie. Gustav Fisher Verlag, Stuttgart.

Thorington, R. W., Jr. 1968. Observations of squirrel monkeys in a Colombian forest. Pages 69–85 *in* L. A. Rosenblum and R. W. Cooper, eds. The squirrel monkey. Academic Press, New York.

MOVEMENTS OF A WILD NIGHT MONKEY (*AOTUS TRIVIRGATUS*)

Richard W. Thorington, Jr., Nancy A. Muckenhirn, *and* **Gene G. Montgomery**

INTRODUCTION

The natural history of the night monkey (*Aotus trivirgatus*) is known in rough outline and is described in Enders (1935), Cabrera and Yepes (1940), and Moynihan (1964). The last presents detailed descriptions of the behavior of night monkeys. These and other papers, however, lack data on the home ranges and activity patterns of *Aotus* that are important for estimating and censusing wild populations. With the increased use of *Aotus* in medical research in the United States (averaging 4,500 animals per year in 1968 and 1969), it is becoming increasingly important to have data that can be applied to proper population management. Although this study was conducted on a near-minimal sample of one animal, we present our data to document (1) the behavior and strategy of an animal released in a strange environment (possibly simulating the dispersal of a young *Aotus* from its parental home range), (2) the movements of a night monkey in a tropical forest, and (3) the feasibility of studying *Aotus* by radio-tracking.

MATERIALS AND METHODS

An 800-g young male *Aotus,* purchased in the local Panamanian market, was fitted with a 40-g dummy transmitter (packaged as a neck collar) and was maintained in a cage approximately $2 \times 2 \times 2$ m for 4 weeks on Barro Colorado Island, C.Z. The animal seemed undisturbed except for occasionally chewing on the whip antenna of the dummy transmitter and regularly holding the antenna with one hand.

On 14 September 1971, when we released the animal, the dummy transmitter was replaced with a 37 g radio transmitter (AVM Inst. Co., Champaign, Ill., Model ST-1) with a predicted battery life of 85 days. The collar, made of rubber-coated, test-lead wire, 4 mm in diameter, acted as a transmitting antenna. We used two harmonics of the basic transmitter frequency (Montgomery *et al.,* 1973) at about 150 MHz and 450 MHz for locating the animal. A 10.2-cm-long whip antenna of the rubber-coated wire increased transmitter output at the higher frequency.

To locate the *Aotus* we used a hand-held Model LA-11-5 receiver with an auxiliary 450-MHz converter (AVM Inst. Co., Champaign, Ill.) with two different antennas. A two-element yagi was used to receive the stronger 150-MHz signal from up to 1.2 km away from the transmitter. A five-element yagi antenna was used to locate the transmitter precisely from close range with the weaker 450-MHz signal (Montgomery *et al.,* 1973).

The activity pattern of the animal was determined by recording changes in intensity of signal from the transmitter with a Rustrack model 291 recorder (Inst. Control Co., Minneapolis, Minn.). Fluctuations in intensity attributable to activity of the animal (Sunquist and Montgomery, 1973) and direct observations of the animal's activity were summed in 5-min increments for each hour of the night to produce estimates of the percentage of time during which the *Aotus* was active or inactive.

Visual observations were occasionally possible with the aid of headlamps and binoculars 7×35 mm. We also used a Starlight scope, but this was not as useful because of the lower magnification (4 ×) of the Star-light scope, the backlighting of the monkey by star-light, and the shallow depth of field of the scope.

RESULTS

The night monkey was released at the site of its cage on 14 September 1971. We located the animal irregularly during the first three nights (a total of 15 radio-locations) and then followed it almost constantly (approximately 80 hours) for the following nine nights.

During the first night after release, the animal moved approximately 120 m northwest and found a vine-covered tree, which served as its home tree for the next 6 days. After the seventh night, it used another tree 45 m west of the first and two other trees within 15 m of the second (Figure 1). These home trees served as daytime refuges for the next 6 days. All four trees were covered with masses of vines in which the animal concealed itself. The first home tree was large, approximately 25–30 m tall, and contained a den. The other three were small, 10–15 m high, and did not contain dens.

Travels of the animal from its home trees alternated between feeding sessions in the vicinity of the original home tree and exploratory excursions from the vicinity. On the third, fourth, eighth, and tenth nights, it was not found outside a 30- × 20-m area centered west of the first home tree. On the second night the animal moved 200 m north of the home trees, while on the fifth and sixth nights it returned to an area near its holding cage. On the ninth night the animal moved 150 m south; on the eleventh night, 150 m south and 250 m west. In returning to home trees from these explora-

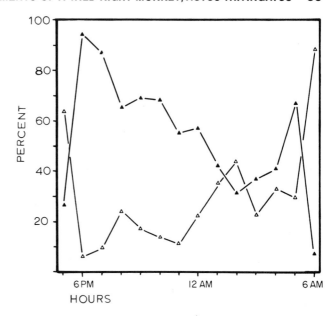

FIGURE 2 Nocturnal activity of *Aotus*. The average percentages of activity (▲) and inactivity (△) per hour are given for the period 14–26 September 1971. Percentages do not total 100 percent for each hour because activity levels could not always be determined.

tory excursions, the animal approximately retraced its original path through the forest canopy. These routes and the areas most used by the *Aotus* are shown in Figure 1.

The animal spent daylight hours 10–15 m above the ground. During the night, some of its pathways brought it closer to the ground (as low as 3 m), but much of its activity took place in the forest canopy. Arboreal pathways used by the animal included very large to very small branches. It moved through the trees quietly, seldom making long jumps. When crossing between trees over small branches, it used its tail extensively for balance.

When feeding on small terminal fruits, such as those of *Brosimum bernadette,* the *Aotus* stretched out to the ends of branches, collected a fruit, sat on a more stable perch to eat it, then repeated the procedure. When feeding on other undetermined foods, perhaps buds and insects, it moved about almost constantly.

The nocturnal activity patterns of the *Aotus* are summarized in Figure 2. The animal became active at dusk between 1700 and 1815 hours and remained active, feeding and moving for several hours. Usually it became inactive after midnight for an hour or two, then increasingly active toward dawn. On at least four of the mornings, it fed intensively for one-half to a full hour before returning to its home tree. It usually retired to a daytime refuge between 0550 and 0600 hours. However, based on Rustrack recordings, it was

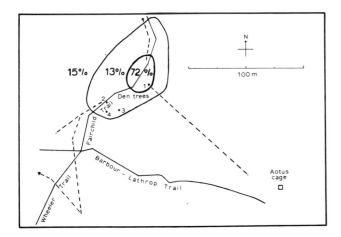

FIGURE 1 Study area on Barro Colorado Island. Percentages indicate relative amounts of time the *Aotus* spent at night within and without the encircled areas. Broken lines indicate routes described in the text.

evident that the animal was active in or near its home tree.

CONCLUSION

When released, the strategy of the *Aotus* appeared to be locating a good home tree with a good feeding area nearby. The animal explored outward from this place and subsequently changed home trees. However, it did not change its principal feeding area during the course of the study. It is unfortunate that there are no similar data on dispersal of other primates for comparison of strategies.

As shown in Figure 1, the *Aotus* spent 72 percent of its time in an area of 800 m² and 85 percent of its time in an area of one-half hectare. The latter area is slightly smaller than the territories of two to four *Callicebus*, as described in Mason's (1966) study. Further studies will be required to determine if our animal occupied a home range typical for *Aotus* in this habitat and if the size of the home range varies in different habitats.

As we radio-tracked the animal, there were long periods during which we could not see or hear it. It moved quietly through the trees, did not vocalize, and usually did not drop rinds or other debris as it fed. With a headlamp we occasionally saw its eyeshine. The animal could not have been followed without use of the radio equipment. Based on our experience with this animal, we feel that it would be feasible to conduct a detailed ecological study of wild *Aotus* by using radio-telemetry and probably impossible to conduct such a study without this field technique.

ACKNOWLEDGMENTS

This study was supported by Air Force Contract No. F44620-67-C-0063 to the Smithsonian Institution (RWT and NAM) and by Smithsonian Research Foundation Grant No. SG 3540001 (GGM). W. W. Cochran designed the two-frequency radio-location system.

The transmitter was purchased by Animal Resources Branch, NIH, USPHS contract PH-43-64-44 (Task Order 12) and Headquarters U.S. Army Medical Research and Development Command Contract No. DADA 17-71-C-1117 to the National Academy of Sciences.

REFERENCES

Cabrera, A., and J. Yepes. 1940. Historia natural Ediar; mamiferos sub-americanos. Buenos Aires.

Enders, R. K. 1935. Mammalian life histories from Barro Colorado Island, Panama. Bull. Mus. Comp. Zool. 78:385–502.

Mason, W. A. 1966. Social organization of the South American monkey, *Callicebus moloch:* a preliminary report. Tulane Stud. Zool. 13:23–28.

Montgomery, G. G., A. S. Rand, and M. E. Sunquist. 1973. Postnesting movements of iguanas from a nesting aggregation. Copeia 1973:620–622.

Moynihan, M. 1964. Some behavior patterns of platyrrhine monkeys. I. The night monkey (*Aotus trivirgatus*). Smithson. Misc. Coll. 146(5):iv, 1–84.

Sunquist, M. E., and G. G. Montgomery. 1973. Activity patterns of two- and three-toed sloths. J. Mammal. 54:946–954.

THE NONHUMAN PRIMATES OF COLOMBIA

Jorge Hernández-Camacho *and* Robert W. Cooper

INTRODUCTION

The faunal, floral, and physiographic diversity of Colombia is among the richest of the neotropics. Although many areas of the country are not as yet well studied, at least 12 genera and 22 species of nonhuman primates are presently known. Only Brazil, with about 16 genera and 32 species and 6.5 times more land area, has more nonhuman primate forms.

It is not possible, based on presently available data, to say exactly where the suborder Platyrrhini first arose. However, the origin of primates in Colombia probably dates to the Tertiary period, when the first platyrrhines became established in South America. Of the five known families—Callitrichidae, Callimiconidae, Cebidae, Homunculidae (known from Argentina), and Xenothricidae (known from Jamaica)—only the latter two are absent in the fossil and/or recent record of Colombia. From the middle Magdalena Valley three monotypic genera of Cebidae (*Cebupithecia, Neosaimiri,* and *Stirtonia*) are known from the late Miocene period (Hershkovitz, 1970). Even so the fossil record for neotropical primates is very limited at present, with no known material available from Mexico and Central America and with the earliest South American records being those from the Tertiary period in Colombia, Bolivia, and Argentina.

In historic times nonhuman primates were first documented in Colombia in some of the early chronicles following the Spanish conquest. Cotton-topped tamarins (*Saguinus oedipus*) and white-throated capuchins (*Cebus capucinus*) from Colombia and tufted

capuchins (*Cebus apella*), squirrel monkeys (*Saimiri sciureus*), and black spider monkeys (*Ateles paniscus*) probably from the Guianas were first described by Linnaeus in 1758 based on several such early accounts. By 1766 Linnaeus had also described red howler monkeys (*Alouatta seniculus*) from Cartagena based on a seventeenth-century account of von Jacquin. In 1807 Count von Hoffmansegg first described, from the Pará region of Brazil, the two species of titis (*Callicebus moloch* and *C. torquatus*) that also occur in Colombia. The next major advance in descriptive knowledge of primates to occur in Colombia came with the travels of the famous German explorer Alexander von Humboldt in the early nineteenth century. He first reported the existence of night monkeys (*Aotus trivirgatus*), black-headed uakaris (*Cacajao melanocephalus*), woolly monkeys (*Lagothrix lagotricha*), and white-fronted capuchins (*Cebus albifrons*). At about the same time (1812) Etienne Geoffroy-St. Hilaire described monk sakis (*Pithecia monachus*), probably based on specimens from Brazil. In 1823 von Spix, after 6 years in the Amazon region, reported the existence of pygmy marmosets (*Cebuella pygmaea*) and both saddle-backed and black-mantled, white-lipped tamarins (*Saguinus fuscicollis* and *S. nigricollis*). By 1823, 11 of the 12 genera and 16 of the 22 species known in Colombia had been described. The genus *Callimico* and five additional species (*Alouatta palliata, Saguinus geoffroyi, S. leucopus, S. graellsi,* and *S. inustus*) were then reported over the next 128 years, the latest being *S. inustus* in 1951.

The development of descriptive, zoogeographic, and

35

FIGURE 1 Distribution map of *Cebuella pygmaea* in Colombia.

natural historical information on neotropical primates has been relatively slow and sporadic, and the information produced in most areas of inquiry has been rather superficial. Unfortunately, this chapter will seldom depart far from that tradition. However, by summarizing some historical information, as well as more recent data based largely on the senior author's diverse field and museum experience in Colombia, we have attempted to add a few more bricks and possibly some missing mortar to the growing foundation of Colombian and general neotropical primate knowledge. Proceeding by approximate phylogenetic order of species, regional common names will be given, as well as known or probable distribution patterns, taxonomic notes and suggestions, general habitat preferences, some aspects of natural history, probable present population status, and important adverse population pressures. Our intention is to provide points of reference and departure for students of primate biology in Colombia. By summarizing much of the basic information presently available to us, we hope to stimulate additions and corrections from the likely wealth of obscure published and unpublished information presently known only to a few. In this spirit, we welcome constructive criticisms, new data, suggestions, etc., directed either to the authors and/or published for the benefit of all.

1. *Cebuella pygmaea* (Spix, 1823)—Pygmy marmosets.

COMMON NAMES: "Chichico" along the Putumayo River; "Leoncito" in the Putumayo and Amazon regions; and "Piel Roja" and "Mico de Bolsillo," popularized by animal dealers.

DISTRIBUTION: (Figure 1) *Cebuella* is well known in the Colombian Amazon region south of the Caquetá River. Reports of its existence further north, in the upper Guaviare River region, remain to be documented. A captive specimen reportedly from Caño Morrocoy, on the south bank of the Guayabero River, about 2 km eastward from the town of El Refugio (also known as La Macarena), is the best evidence to date.

In Colombia *Cebuella* is typically an inhabitant of mature, nonflooding forest. Its ecological association with the guarango tree (*Parkia* sp.) as a source of sap is one of the more striking dietary specializations known among neotropical primates.

HABITS: Pygmy marmosets are rather difficult to find due to their small size, the camouflage of their coat color, their squirrellike habit of moving to the opposite side of a trunk when disturbed, and their lack of any conspicuous physical or vocal display. They have been observed in groups as large as 10 or 15 and

are seldom, if ever, seen alone. They are most often found on the trunk or major branches of a guarango tree and, if disturbed while low on the trunk, are known to run to the ground for escape. Their terrestrial route to and from guarangos is evident by their intermittent presence in such trees in clearings a short distance from undisturbed forest. As mature guarangos are usually emergent over the forest canopy and somewhat free of climbers, it appears that such terrestrial passage must be rather common. The stomach contents of specimens examined in March 1965 in the vicinity of Puerto Leguízamo, Comisaría of Putumayo, contained largely jellylike, dirty-whitish-colored guarango sap in addition to some finely crushed insects (mostly Coleoptera) and evidence of fruit pulp. The small sap-producing holes in the bark of the trunk or branches of guarango used by *Cebuella* appear to be produced largely, if not exclusively, by their procumbent lower incisors.

STATUS: As in many other Colombian primate species, the exact status of *Cebuella* is uncertain. It is fortunate that guarango is not a highly desirable timber species and is large enough to be left standing in many man-made clearings. Pygmy marmosets, of course, are unlikely to be hunted for their meat, but are sometimes collected for commercial purposes. In some areas the centuries old method of ringing a guarango trunk with a sticky resin to capture pygmy marmosets as they enter or leave the tree is used with considerable success. It is likely that this means of trapping is as ancient as the practice of some upper Putumayo River Indians who keep captive *Cebuella* to pick lice from their hair. *Cebuella* is also captured on occasion in banana-baited traps (intended for *Saguinus* spp.) near ground level. It is doubtful that such limited commercialization is nearly as threatening to the relatively prolific *Cebuella* as is habitat destruction. However, as discussed elsewhere in this volume (Moynihan, 1976), *Cebuella* seems to have remarkable adaptive abilities with regard to living as a near commensal of man in highly degraded habitat situations.

2. *Saguinus nigricollis* (Spix, 1823)—White-lipped, black-mantled tamarins.

COMMON NAMES: "Leoncito" in the upper Putumayo River region; "Bebeleche" in Amazonian Colombia.

DISTRIBUTION: (Figure 2) In Colombia *S. nigricollis* occurs within about the same region as *Cebuella pygmaea* with the Caquetá River forming the northern limit. The population of the upper Putumayo River has a dull and brownish cast to the lower back and hind limbs, as well as some grizzled yellow and black in the saddle. This population is thus more reminiscent of *S.*

FIGURE 2   **Distribution map of the genus *Saguinus* (part) in Colombia.**

fuscicollis than is the lower Putumayo and Leticia population, which has a rich ferrugineous cast to the lower back and hind limbs and no yellowish tones in the saddle area. Our material is inadequate to determine whether or not these differences represent subspecific distinctions. *S. nigricollis* in Colombia occupies a wide range of rain forest habitats, from primary to secondary growth, and includes both seasonally flooded and nonflooding areas.

HABITS: Group size seems to average between 5 and 10 individuals, whose high-pitched vocalizations are often heard well before the group is seen by an observer. They seem to prefer a middle canopy level and are most often seen at about 8 or 10 m above the forest floor. Stomach contents usually contain a mixture of fruits, berries, and insects. To date the only interspecific associations known to us are observations of apparently mixed groups of *S. nigricollis* with *S. fuscicollis* in the area of Puerto Leguízamo.

STATUS: For the past few years *S. nigricollis* has been captured in the Leticia region for use in biomedical investigation (principally virus cancer and infectious hepatitis studies). *S. nigricollis* was the most frequently encountered primate species during a field survey conducted in March 1972 from Leticia to a point some 70 km upriver on the Colombian bank. Within a 1,000-km² region fronting on the Amazon and extending 15 to 18 km from its north bank, this species was encountered 10 times in a total of 75 hours of foot travel in undisturbed rain forest. It seems to be a rather adaptive species, living both close to human habitation and plantations and deep within relatively undisturbed primary forest.

3. *Saguinus graellsi* (Jiménez de la Espada, 1870).

COMMON NAMES: "Bebeleche" in the Putumayo region.

DISTRIBUTION: (Figure 2) *S. graellsi* is known in Colombia from the Comisaría of Putumayo on the basis of a preserved specimen (Universidad Nacional de Colombia, Bogotá), a number of reliable sightings (e.g., Moynihan, 1976), and captive specimens from the neighborhood of Puerto Asís eastward to the vicinity of Puerto Leguízamo. The northern limit is probably the southern bank of the Caquetá River. The eastern limit is unknown. *S. graellsi* is undoubtedly sympatric with *S. fuscicollis fuscus* throughout its range as well as with the population of *S. nigricollis* in the region of Puerto Leguízamo.

HABITS: Unknown.
STATUS: Unknown.

4. *Saguinus fuscicollis* (Spix, 1823)—White-lipped, saddle-backed tamarins.

COMMON NAMES: "Bebeleche" throughout its range.

DISTRIBUTION: (Figure 3) *S. fuscicollis* inhabits forested lowlands of the Amazon Basin in Colombia southward from the Caquetá River, northward in the Intendencia of Caquetá (eastward at least to the western bank of the Yarí River and northward to the southern bank of the Guayabero River in the southern Department of Meta), and eastward to the region around San José de Guaviare on the southern bank of the Guaviare River in the Comisaría of Vaupés. The species ranges upward to 500 m in the Andean piedmont.

The populations of the Putumayo River, including one specimen from the right bank of the Guayabero River in Angostura near the southern tip of the Macarena Mountains, are referrable to *S. fuscicollis fuscus* (Lesson, 1840). Unpreserved specimens from San José de Guaviare suggest that a presently undescribed subspecies inhabits this region. Live specimens recently examined in Leticia and reportedly from Puerto Narino on the Colombia bank of the Amazon are referrable to *S. fuscicollis triparitus* (Milne-Edwards, 1878). *S. fuscicollis* is not known in the immediate region of Leticia. Also, no specimens are available for the region between the Guamués and the Sucumbíos Rivers of the southwestern Comisaría of Putumayo.

HABITS: *S. fuscicollis* is usually encountered in groups of 5–20 individuals in primary forest, advanced second growth, or even areas of human activity. Stomach contents indicate a preference for various berries and insects. *S. fuscicollis* has been observed in mixed groups with *S. nigricollis* in the region of Puerto Leguízamo.

STATUS: *S. fuscicollis* is not seriously exploited in Colombia and not well known in all areas of its supposed range. Its population status is probably unthreatened and it may well have a somewhat interrupted distribution.

5. *Saguinus inustus* (Schwartz, 1951).

COMMON NAMES: "Mico Diablo," "Diablito," and "Tití Diablito" in the Comisaría of Vaupés.

DISTRIBUTION: (Figure 3) At present the species is known in Colombia from authentic records in the area of Mitú on both sides of the Vaupés River, as well as from San José Guaviare, on the southern bank of the Guaviare River, about 1 hour by boat downstream from the confluence of the Guayabero and Ariari rivers. Scattered reliable reports indicate its presence in other areas of the Comisaría of Vaupés and Guainía (Inírida and lower Guaviare rivers). It is likely that the Guaviare River forms the northern limit of the species, possibly as far west as some point between the towns

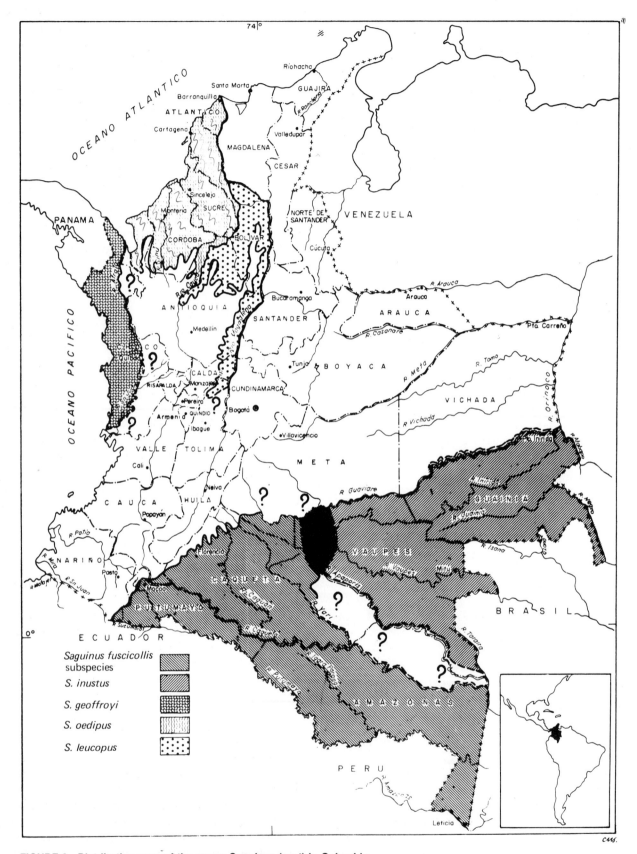

FIGURE 3 Distribution map of the genus *Saguinus* (part) in Colombia.

of El Refugio and San José Guaviare; in 1959 the species was unknown as far north as the immediate vicinity of El Refugio. The southern limit is probably the Apaporis River, based in part on its history as an effective barrier for a number of other species and subspecies of terrestrial and arboreal vertebrates. It is also possible that its range may extend northward to the gallery forests of the Ariari from accounts of recent settlers living near Granada.

HABITS: Although the species is found in rain forest in small groups, it is totally unstudied in Colombia. The possibility exists that its range is interrupted due to the heterogeneity of vegetational types (including chersophytic and chasmophytic dwarf forests, scrubs, and savannas) in Vaupés and Guainía.

STATUS: Probably unthreatened but not known with certainty.

6. *Saguinus geoffroyi* (Pucheran, 1845)—Geoffroy's tamarins.

COMMON NAMES: "Titi" and "Bichichi" in the Department of Chocó.

DISTRIBUTION: (Figure 3) *S. geoffroyi* occurs in Colombia from the Panamanian border probably as far south as the San Juan River, on the Pacific coast. The eastern limit is the western bank of the Atrato River. The upper altitudinal limit of its range is unknown, but it probably does not exist above 800 m of elevation. It seems more abundant in secondary forests or early secondary growth interspersed with plantations and seldom inhabits deep climax forest. Of the few museum specimens of *S. geoffroyi* known for Colombia as well as those seen in captivity (largely from the region of Acandí), a large percentage have distinct sulfur-yellowish underparts, including lightly pigmented areas of the limbs. This characteristic is seen in both juveniles and adults of either sex and appears to be individually variable, i.e., other specimens from the same region have perfectly white underparts, hands, and feet.

HABITS: See *S. leucopus*.

STATUS: *S. geoffroyi* is very poorly known in Colombia due to the relative remoteness of the region in which it occurs. However, it is seldom exploited, and human population is neither very large nor expanding greatly within this area.

7. *Saguinus oedipus* (Linnaeus, 1758)—Pinches or cotton-topped tamarins.

COMMON NAMES: "Titi," "Titi Blanco," "Titi Leoncito," and "Titi Pielroja" throughout the range in northwestern Colombia.

DISTRIBUTION: (Figure 3) *S. oedipus* is endemic to Colombia and occurs from the Urabá region of the northwestern Department of Antioquia southward at least to the Leon River and in the departments of Córdoba, Sucre, northern Bolívar, and Atlántico. The eastern limit is the western bank of the lower Magdalena and lower Cauca rivers, extending into north-central Antioquia. Recent investigation has disclosed that *S. oedipus* is absent from Mompós Island and is there replaced by *S. leucopus*. Cotton-topped tamarins occur in rain forest, deciduous forest, and second growth but are absent from xerophytic forest. Their highest reported localities are at elevations of about 400 m but they may range somewhat higher in the upper Sinú Valley. The vast majority of *S. oedipus* appear to have totally white underparts, forearms, hands, and feet. Only one museum specimen and several captive specimens from the San Jorge River basin have been observed to possess a sulfur-yellowish tinge to these areas. Another seemingly individually variable character is the shade and extent of the rufous coloration of the thighs and proximal tail.

HABITS: See *S. leucopus*.

STATUS: *S. oedipus* has been heavily commercialized for at least 10–15 years. It is likely that as many as 30,000 to 40,000 cotton-topped tamarins have been exported during this period. Much of the species' habitat lies within a major cattle-grazing area of Colombia in which suitable forests have continually been cut or degraded. The present population status of *S. oedipus* is unknown in absolute terms; however, concern for its survival is growing. Although the species is able to exist in small forest remnants, second growth, and in greatly altered areas, so much of its original habitat has been effectively destroyed that *S. oedipus* has been protected by law in Colombia since 1969.

8. *Saguinus leucopus* (Günther, 1876)—White-footed tamarins.

COMMON NAMES: "Titi" and "Titi Gris" throughout the range in northcentral Colombia.

DISTRIBUTION: (Figure 3) *S. leucopus* is endemic to Colombia and occurs in northeastern Antioquia (regions of Cáceres, Valdivia, and the Nechí River Valley); southern Bolívar, including Mompós Island; and the western bank of the middle Magdalena River in the departments of Antioquia, Caldas, and northern Tolima (at least as far south as the vicinity of Mariquita). Its general limits are the eastern bank of the lower Cauca River, the western bank of the middle Magdalena River (including all of the larger river islands), and the foothills of the central Andes. Coloration in *S. leucopus* seems to be rather constant, with the known exception of specimens from the extreme southern range of Parroquia de Bocaneme near Mariquita, whose basal portions of all colored hairs are decidedly darker brown than observed elsewhere. In addition

FIGURE 4   Distribution map of *Callimico goeldii* in Colombia.

there are distinct dusky ulnar and tibial stripes and the tail is black rather than brown (except for the white tail tip). Specimens from several regions also demonstrate an individually variable, brown, elongated spot over the fourth metatarsal of each foot.

HABITS: On the basis of scant information available to us, no real distinction can be made between the habits and behavior of *S. geoffroyi*, *S. oedipus*, and *S. leucopus*. Group size reportedly ranges from 3 to 12 or more individuals. They are found at all levels of the forest and most frequently at its fringes near streams or other natural barriers and in areas of second growth. Food habits are diverse but include a strong preference for insects and fruits (particularly various types of berries found in the understory or in second growth). Also, on one occasion an adult male *S. oedipus* in a semicaptive environment was observed to leap upon, decapitate with one bite, and consume in total a small *Iguana iguana*. It seems possible that the tendency of these species to be active in areas with more direct solar exposure may be related in part to their taste for insects and small lizards. Our observations in Colombia also lend support to the suggestion (Moynihan, 1970) that the decreasing abundance of *S. geoffroyi* on Barro Colorado Island in the Panama Canal Zone is probably related to the elimination of important ecological niches as the forest reaches an advanced successional stage.

All of these closely related tamarins react in apparent fright to raptorial birds, including even rather small species. Interaction with other primate species is unknown to us, although all three forms live in or near habitat often occupied by *Cebus capucinus* or *C. albifrons*. Data on possible seasonal reproduction in the wild are limited, but it seems that such a tendency exists, based both on records in captivity (Cooper, unpublished data) and unconfirmed field reports. Although twin births are characteristic of tamarins and marmosets in general, people familiar with these species in the wild are seldom aware of this condition from field observation.

STATUS: *S. leucopus* is little exploited, but its habitat has been greatly reduced due to clearing of forests, particularly during the past 20 years.

9. *Callimico goeldii* (Thomas, 1904)—Goeldi's marmosets.

COMMON NAMES: None in Colombia.

DISTRIBUTION: (Figure 4) In Colombia *C. goeldii* has been collected in two localities between the upper Putumayo and Caquetá rivers (Hernández-Camacho and Barriga-Bonilla, 1966) and from a third locality in the lower Guamués River, a major southern tributary of the upper Putumayo. The northwesternmost collection

was on the Igara-Paraná River near La Chorrera; and the intermediate locality is Quebrada del Hacha, on the north bank of the Putumayo River. All specimens have been collected in nonflooding forest, either level or with low rolling hills.

HABITS: According to the Kofán Indians in the Guamués region, *Callimico* is found most frequently in the understory and even on the ground, and its diet includes berries and small fruits. The single specimen collected near the mouth of the Quebrada del Hacha was associated with a group of *Saguinus fuscicollis fuscus* and was the only *Callimico* seen.

STATUS: The species is poorly known over most of its supposed range in Colombia. The status of any local population is unknown. Occasional specimens that have been seen in the Leticia market are thought to be from upriver as far as the Ucayali, in Peru.

10. *Saimiri sciureus* (Linnaeus, 1758)—Squirrel monkeys.

COMMON NAME: "Titi" in the eastern plains, the Caquetá region, and the upper Magdalena Valley; "Vizcaino" in the Caquetá and the upper Guayabero River regions; "Mico Fraile," "Fraile," and "Frailecito" in the Putumayo River and Leticia regions; "Barizo" on occasion in the Leticia area (of Peruvian origin); "Saimiri" in the Leticia region (from Tupi Indian roots); "Menechino" (Tukano Indians, *fide* Olalla from specimen labels).

DISTRIBUTION: (Figure 5) In Colombia *Saimiri* occupies all of the Colombian Amazon and the piedmont of the eastern Andes and a considerable portion of the southeastern plains. The limit of its northeastern extension is not well defined, but it appears that increasing grasslands and decreasing gallery forest and rainfall are limiting factors. It occupies a wide range of forest types from gallery to low canopy sclerophyllous and hillside forests to palm forests (particularly associations of *Mauritia flexuosa*) and both seasonally flooded and nonflooding rain forests. In the upper Magdalena Valley it has a similar distribution to *Cebus apella*, and its northernmost range is not well defined. The maximum elevation at which it has been observed is about 1500 m in this region. The llanos (eastern plains) and piedmont populations southward to the Caquetá River are rather uniform in appearance and differ from those of the Colombian Amazon mainly in having grizzled gray rather than yellow forearms and wrists. Their subspecific designation is clearly *S. s. caquetensis* (J. A. Allen, 1916) but the subspecific status of the Amazon populations is awaiting careful revision of the entire species (see Cooper, 1968).

HABITS: Group size in the gallery forests of the

FIGURE 5 Distribution map of *Saimiri sciureus* in Colombia.

eastern plains is relatively small, as noted by Thorington (1968) and by Baldwin (1971). Apparent single individuals are seen, and a maximum of about 30, mostly subadult, specimens was observed by Hernández-Camacho traveling with a group of *Cebus apella* in piedmont forest near Villavicencio in June 1955. Group size in the Colombian Amazon is typically larger, but we are not aware of reliable accounts of more than 40 or 50 individuals. In all habitat types they seem to travel mostly beneath the highest available canopy but have been observed to come to the ground on rare occasions. Their food preferences are very broad and include fruits (e.g., *Cecropia* spp., *Ficus* spp., *Euterpe* spp., Rubiaceae, *Campomanesia* sp., etc.), berries, insects, spiders, etc. Thorington (1968) noticed births in the eastern plains in February or March, and Amazon populations seem to have about the same birth season (Cooper, 1968).

STATUS: Squirrel monkeys are not seriously hunted for food in Colombia, but they are captured commercially in large numbers in the Leticia region. In March 1972 only one *Saimiri* group was encountered in nearly 75 hours of travel in undisturbed rain forest along the Colombian Amazon from Leticia upriver some 70 km. Although systematically collected data are lacking, it seems likely that numbers may be decreasing in local areas of intensive trapping and/or habitat destruction. The upper Magdalena Valley population is the most seriously endangered due to habitat destruction, and protective measures are definitely needed.

11. *Aotus trivirgatus* (Humboldt, 1811)—Night monkeys, owl monkeys, or Douroucoulis.

COMMON NAMES: ''Marta,'' ''Martica,'' and ''Marteja'' in northern Colombia; ''Marta'' in the departments of Antioquia, Caldas, Quindio, Risaralda, Tolima, and northern Valle; ''Mico Dormilón'' in central Colombia; ''Tutamono'' and ''Tutumono'' in Meta and the Amazon; ''Sorbehumo'' in Meta; ''Mico de Noche'' in central Colombia and the Amazon; ''Maco Cagao'' in Santander. In general, except for ''Maco Cagao,'' ''Mico de Noche,'' ''Mico Dormilón,'' and ''Sorbehumo,'' these names are also used for kinkajous (*Potos flavus*), olingos (*Bassaricyon gabbii*), woolly opossums (*Caluromys* spp.) and pygmy anteaters (*Cyclopes didactylus*). ''Sorbehumo'' is also applied to hawks like *Buteo platypterus* and *Buteo swainsoni* due to their habit of hunting at the edge of a fire (*Aotus* is said to approach campfires at night). The name ''Douroucouli'' is an onomatopoeic rendition of the vocalization of *Aotus* based upon the report of Justin Goudot (quoted by Alston, 1879).

DISTRIBUTION: (Figure 6). *Aotus* is known in all of Colombia except for the Guajira Desert in the northeast, the altitudinal zones over 3,200 m, the northeastern plains, and several local savanna-and-scrub mountain areas of the comisarías of Guainía and Vaupés in the geological zone known as the Guayana Shield. With the exception of mangrove swamps, *Aotus* typically inhabits every major forest zone of Colombia, including second growth and even well-shaded coffee plantations. *Aotus trivirgatus* subspecies, found throughout Colombia, may best be described according to relatively distinct regional characteristics of color pattern on the hands and feet and length of hair coat.

A. t. griseimembra (Elliot, 1913) occurs in the lowlands of northern Colombia as well as the Santa Marta Mountains and extends westward to the neighborhood of the Sinú River, the San Jorge River, the lower Cauca River, and the lowlands of the middle and upper Magdalena River Valley. The subspecies is typically characterized by a relatively short and somewhat adpressed hair coat, upperparts with a decidedly brownish or yellowish-brown cast, underparts rather dull-to-light yellowish, and the dorsal surfaces of hands and feet light brownish with rather inconspicuously dark hair tips. Unfortunately, the type specimen with a strong dark admixture on the hands, and black sides on the feet, is exceptional; an exact topotype fits the more usual pattern as described previously. The only other specimens seen at any significant variance with this general description are those of the Santa Marta Mountains at higher elevations (which have a tendency toward longer hair coats) and a single adult male from Ayacucho near La Gloria (500 ft) in the middle Magdalena Valley (which had very prominently dark hair tips on the dorsal hands and feet). Also, an adult pair and a subadult female from Sierra Negra (1000 m) in the Perijá Mountains east of the Santa Marta Mountains near the Venezuelan border had longer hair coats, but the distal portion of the tail was dark brown instead of black, differing from all other known populations of *A. trivirgatus*.

A. t. zonalis (Goldman, 1912) occurs in Colombia from the Panamanian border in the Pacific lowlands southward at least to the Raposo River just south of Buenaventura, the Urabá region and eastward to the Sinú Valley, and possibly through the upper San Jorge Valley to the Puerto Valdivia region of northern Antioquia. This subspecies is completely homogeneous throughout its known range in Colombia as well as in eastern and central Panama. It resembles typical *A. t. griseimembra* in all respects except that the dorsal hair of its hands and feet appears dark brown or blackish. We do not relegate *A. t. zonalis* to a synonym of *A. t. griseimembra* as proposed by Hershkovitz (1949),

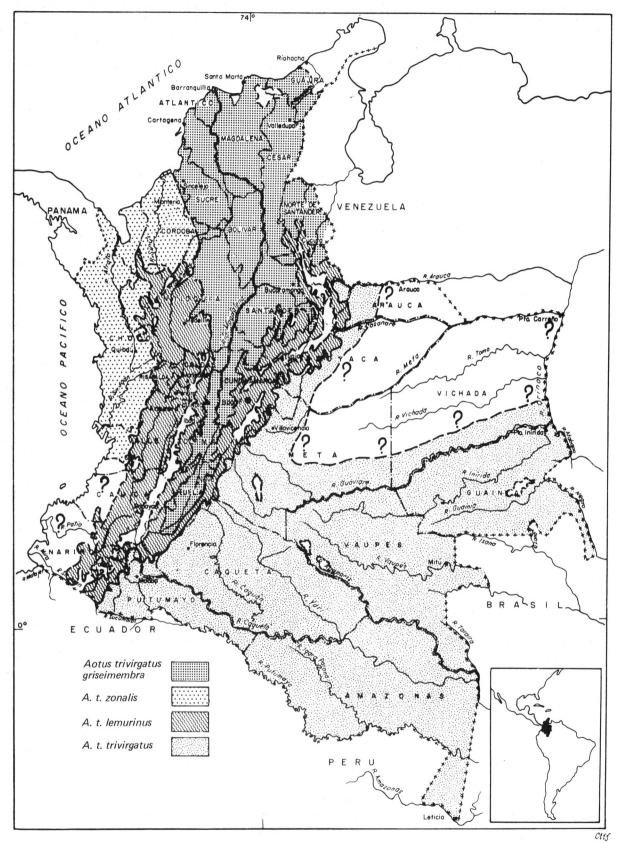

FIGURE 6 Distribution map of *Aotus trivirgatus* in Colombia.

because our examinations of literally hundreds of living specimens of the latter from the Barranquilla market have revealed only specimens with essentially light-colored hands and feet.

A. t. lemurinus (I. Geoffroy-St. Hilaire, 1843) occurs in the three Andean mountain chains of Colombia at elevations from about 1,000–1,500 m upward to the timber line (3,000–3,200 m). It is a rather variable subspecies appearing quite often with two basic color phases, which can both be found in the same family group. One is decidedly grayish-brown, and the other is a richer, more reddish-brown in the upper parts. However, a range of intermediate coloration can be found. The underparts are always a rather dull yellow, indistinguishable from *A. t. griseimembra* and *A. t. zonalis*. The hair coat is extremely long and soft and is the most valuable distinguishing characteristic. The hands and feet of this species are remarkable in their color variation and lack of full correlation, even in individual specimens. Specimens examined from the western Andes (around Cali) and most of the specimens from the central Andes have black-tipped hairs on the hands and feet (at least reaching the distal carpus and tarsus). A few specimens from the western Andes and a number of those from the central Andes appear variably grizzled in color on the metatarsal and metacarpal regions due to reduced extent of the dark tips, thus allowing exposure of the lighter hair bases. In the eastern Andes a full range of individual variation occurs with regard to this character, i.e., from extensive black hair tips to the very reduced tips typical of *A. t. griseimembra*. Some variation has also been observed between the hands and feet of the same individuals with regard to this characteristic.

A. t. trivirgatus (Humboldt, 1812) occurs in the eastern plains and piedmont at least southward to the Guaviare River and probably extends continuously into the upper Amazon basin. It is characterized by a short-to-medium hair coat usually of a rather pure gray color with comparatively light-colored hands and feet due to rather short, dark, apical hair tips (as a rule only slightly more apparent than those of *A. t. griseimembra*). Although no specimens are available to us from the Guaviare to the Putumayo rivers, those from the Leticia area (as well as selected specimens from northwestern Brazil and southern Venezuela) have more richly golden-orange underparts in contrast with the more typically dull and lighter yellow underparts of the specimens from the eastern plains and piedmont. Such specimens are also consistent with the characters claimed by Calvalho for the recognition of *A. t. vociferans* (Spix, 1823) from the Solimoes region (type locality Tabatinga, Brazil). However, more

material from intermediate areas will be necessary to clarify the situation.

It is noteworthy that in all subspecies of Colombian *A. trivirgatus*, coloration patterns of the head were found to be variable and totally unreliable taxonomic characteristics.

HABITS: *Aotus* is typically encountered in pairs with one or two subadult offspring. On occasion associations of several adult pairs have been collected in the same nest. The nest sites of *Aotus* are located in tree hollows and/or in woody climbing vines in an accumulation of dry leaves and twigs. Such nests appear identical with those used by *Potos flavus*. Aside from the grossly similar colors and appearance of night monkeys, kinkajous, and olingos, the fact that they may all be encountered feeding in the same tree undoubtedly contributes to local confusion regarding their identities. *Aotus* become active just before dusk and have usually been observed in the field before midnight. This observation is not intended to preclude the possibility that another period of activity may begin in the morning before sunrise. In our experience, *Aotus* is seldom detectable from its vocalization. They have been encountered at almost every level beneath the canopy. An old observation of Goudot (in Alston, 1879, p. 14) states that *Aotus* returned regularly to the same area at night looking for guavas (*Psidium guajava*).

STATUS: *Aotus* is presently one of the most successful primate forms in Colombia, due to its wide altitudinal and regional distribution, broad range of habitat (including rather degraded areas), and general lack of exploitation by man. The only region in which it is presently hunted extensively is the delta of the lower Cauca and neighboring middle Magdalena rivers. This trade has developed principally over the past 4 or 5 years in response to a growing biomedical research demand for this particular population of *Aotus* for use in malaria chemotherapy study (Cooper and Hernández-Camacho, 1975). The effects on their local status caused by removal of several thousand specimens annually is not definitively known at this time. However, until reliable data become available on the trapping methods used, the specific areas of exploitation, the biological cycles of *Aotus,* and the size of the regional population, licenses for trapping and exportation will be carefully screened. Of historical interest is the fact that skins of *Aotus,* along with *Lagothrix* and *Alouatta,* have all been used locally as ornamentation on the brow bands of bridles for horses.

12. *Callicebus moloch* (Hoffmannsegg, 1807)—Dusky titis.

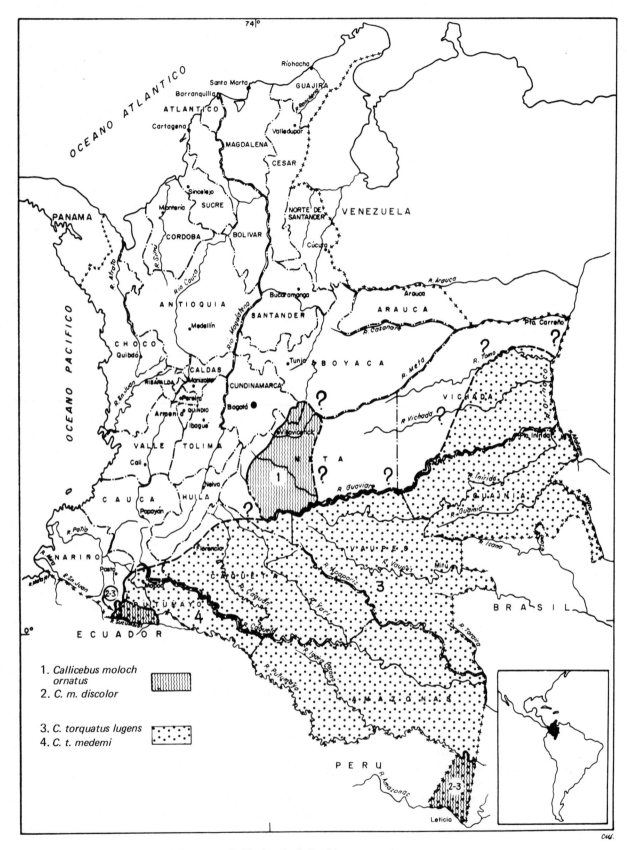

FIGURE 7 Distribution map of the genus *Callicebus* in Colombia.

COMMON NAME: "Socay" or "Zocay" in Meta (*C. m. ornatus*) and in the upper Putumayo River region (*C. m. discolor*), probably of Quechua Indian origin; "Zogui-zogui" in the Amazon (*C. m. discolor*); also "Mico Tocón," of Peruvian origin, in the Leticia area.

DISTRIBUTION: (Figure 7) In the Department of Meta, *C. m. ornatus* (J. E. Gray, 1866) occurs in the piedmont to an upper altitudinal limit of about 500 m as well as in the adjacent lowland gallery forests. It is marginally sympatric with *C. torquatus lugens* along both banks of the Guayabero River at the southern extreme of its range. Its northernmost occurrence is the lower Upía River, a northern tributary of the Meta River. As pointed out by Mason (1966), *C. m. ornatus* is able to survive in even small isolated gallery forests of only a few hectares.

C. m. discolor (I. Geoffroy-St. Hilaire, 1848) occurs in Amazon rain forest and has recently been collected on the southern bank of the Guamués River, a western tributary of the upper Putumayo. Its local existence in Colombia is limited to the region between the Guamués and the Sucumbios River, the next lower western Putumayo tributary. This population is probably continuous with that along the southern bank of the Putumayo River in Ecuador and Peru and also is provisionally equivalent with the population that occurs in the trapezium area of the Amazonas Comisaría of Colombia between the Putumayo and Amazon rivers.

HABITS: Both subspecies are usually encountered in groups of two to four animals. On one occasion (in June 1955) Hernández-Camacho saw a total of seven individuals in a peninsula extending from a larger gallery forest on the Ocoa River, 7 km southeast of Villavicencio. Most sightings have been made in understory trees at heights of 3–8 m, and specimens have been collected on the ground (and crossing savanna in the case of *C. m. ornatus*). *Callicebus* is able to swim when forced and does so with a side-to-side undulation of the body and tail. *C. m. ornatus* often begins vocalization before sunrise, particularly on overcast days. It stops vocalizing somewhat sooner when the mornings are sunny than when they are overcast. They have been observed in sloped piedmont forest with a canopy of no more than 10 m in the southern Macarena Mountains. *C. moloch* subspecies have not to our knowledge been observed in association with other primate species. Specimens of *C. m. ornatus* have produced stomach contents with a predominance of vegetable matter, including berries, seeds, and fruit pulp; leaf parts have never been observed. Also, most specimens (collected between January and March and in June and July) have

included some Coleoptera and Orthoptera, as well as spiders and millipedes.

STATUS: Clearing of level gallery forests and low piedmont areas for agricultural purposes in the Department of Meta northward from the Ariari River during the past 30 years has reduced by more than 50 percent the suitable habitat for this population of *C. m. ornatus*. Human population is continuing to grow in this area due to emigration from mountain regions to the west. However, *Callicebus* is seldom captured commercially or hunted for food in this region. It also displays an unusual ability to survive in isolated forests and in low mountainous areas. Its future is somewhat assured in the Macarena Mountains National Park (800,000 ha). As for the Colombian population of *C. m. discolor*, the region between the Guamués and Sucumbios rivers, now a major Colombian oilfield, is being rapidly settled, while the Colombian trapezium human population is increasing and some limited commercialization of the species is occurring, largely secondarily to food hunting.

13. *Callicebus torquatus* (Hoffmannsegg, 1807)— Widow monkeys and white-handed titis or collared titis.

COMMON NAMES: "Macaco" in the Caquetá and Putumayo regions; "Macaco Caresebo" in the Guayabero River region; "Viuda" and "Viudita" in the Vaupés River and Orinoco River regions.

DISTRIBUTION: (Figure 7) *C. t. lugens* occupies the area from the Guaviare and Guayabero rivers southward to the Putumayo and Colombian Amazon rivers, except for the habitats of *C. t. medemi* in the area between the upper Caquetá and Putumayo rivers. It also exists northward from the lower Guaviare River at least to the lower Tomo River in the Vichada Comisaría. Within this entire area it is found in forest types ranging from gallery forests in the north to uninterrupted rain forests in the Amazonian basin.

C. t. medemi, most remarkable for its black hands, is known only from two localities in the vicinity of Puerto Leguízamo in the upper Putumayo basin. Two *C. torquatus* specimens from the southern bank of the lower Guamués River near Puerto Leguízamo have light-yellowish hands and distinctly ferrugineous underparts (including ventral tail surface). This pattern is in contrast with the blackish underparts of typical *C. t. lugens* and approaches the characteristics claimed for *C. t. lucifer* (Thomas, 1914), a subspecies not separable from *C. t. lugens* according to Hershkovitz (1963). Recent collection (near Leticia) of a *C. torquatus* specimen with rather dull and somewhat brownish upperparts represents the only specimen with

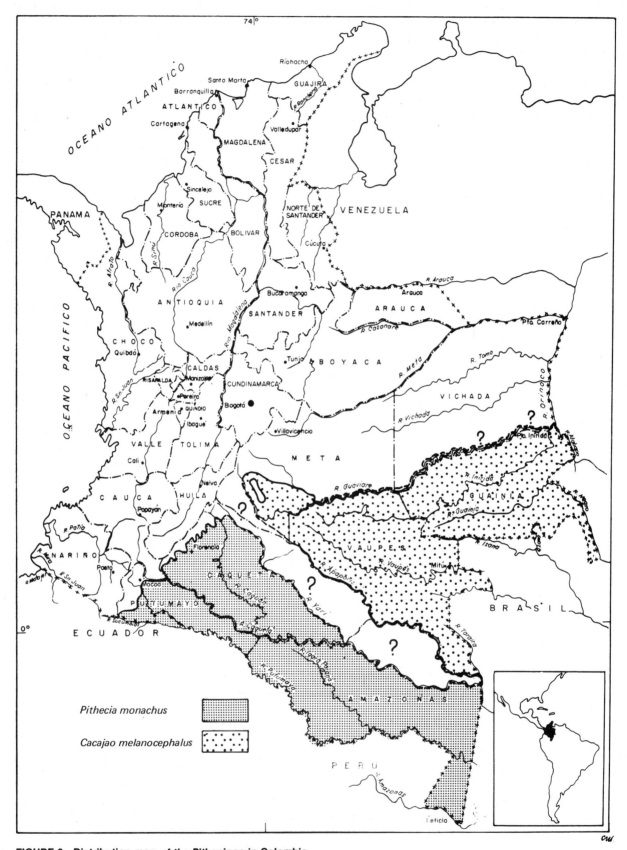

FIGURE 8   Distribution map of the Pithecinae in Colombia.

adequate locality data known to us from the Colombian bank of the Amazon. This description is consistent with brownish individual varients of *C. t. lugens* seen almost throughout the range of the subspecies.

HABITS: Observations have been limited to the northern extreme of *C. t. lugens* distribution (probably also the northernmost limit for the genus) between the Tomo and Tuparro rivers, as well as the area of *C. t. lugens* and *C. m. ornatus* sympatry on the upper Guayabero River. Observed group size has varied from two (Tuparro River) to a maximum of five (Guayabero River), and habitat has ranged from continuous forests to narrow gallery forests. On one occasion a specimen was collected from a moving group of four after sunset (6:30 p.m.) in a wide gallery forest. On another occasion vocalization was heard at about 7:30 p.m. on the Guayabero River. Stomach contents (January to March) contained significantly less arthropod content than observed in *C. m. ornatus* in the same region (Guayabero River). Within the area of sympatry north of the Guayabero, only three groups of *C. t. lugens* were observed over a 3-month study period (as compared to nearly daily sightings of *C. m. ornatus*). On all occasions the *C. t. lugens* sightings were distant from any *C. m. ornatus* sightings. On the southern bank of the Guayabero during the same period, *C. t. lugens* were observed on the average of several times per week, while only one sighting (and collection) of *C. m. ornatus* was made. North of the Guayabero *C. m. ornatus* were observed in January and February to be carrying only older infants (with complete deciduous dentition) on occasions when parent-dependent specimens were seen or collected. During the same period, *C. t. lugens* on the southern bank were observed to be carrying considerably younger infants with adpressed hair and only deciduous incisors erupted in the specimens collected.

STATUS: Considering the much greater range of *C. torquatus* subspecies in Colombia than of *C. moloch* subspecies, the status of the former is of less concern except in the more western areas of heavy settlement. Insofar as food hunting and commercial exploitation are concerned, the population pressures must be considered as not significantly different from those of *C. moloch* subspecies.

14. *Pithecia monachus* (E. Geoffroy-st. Hilaire, 1812) —Monk sakis.

COMMON NAME: "Volador" or "Mico Volador" in Colombia.

DISTRIBUTION: (Figure 8) *P. monachus* (subspecific designation problematic) occurs in the Colombian trapezius north of the Amazon and elsewhere in Amazonian Colombia north of the Putumayo River. Its northwestern limit is well documented as the piedmont of the eastern Andes as far north as the upper Caguán River to an elevation of 500 m. The northeastern extent of its range is uncertain but is at least to the southern bank of the Yarí River and below its mouth to the southern bank of the Caquetá River. Its habitat is always limited to primary, well-developed rain forest.

HABITS: Monk sakis are usually observed as pairs or perhaps small family groups. They are high-canopy dwellers, shy and prone to rapid flight, which has resulted in practically nonexistent data on their behavior and biology. Recent studies of primate ecology in the upper Putumayo River and Caquetá River basins by Kosei Izawa of the Japan Monkey Centre (see Izawa, 1972) will hopefully contribute to a better understanding of this and other primate species of the region.

STATUS: Hunting *Pithecia monachus* for meat in Colombia is not known to us, nor have we seen dust brooms made from its tail as was observed by one of us (R. W. C.) in 1964 in Iquitos, Peru. Specimens, occasionally seen in dealers' compounds in Leticia, are exported from Colombia in very small numbers. Considering its habits and requirements for undisturbed high rain forest, how *Pithecia* might be captured alive is unknown. Its population status in Colombia is undoubtedly better than that of such an obscure species as *Cacajao melanocephalus,* as monk sakis are relatively well known within their range in the Colombian Amazon basin.

15. *Cacajao melanocephalus* (Humboldt, 1812)— Black-headed uakaris.

COMMON NAMES: "Chucuto" in the comisarías of Vaupés and Guainía; "Uacari" for occasional specimens arriving in Leticia (probably from the lower Rio Negro region in Brazil); "Piconturo" (Tukano Indians, *fide* Olalla from specimen labels).

DISTRIBUTION: (Figure 8) Distribution is very poorly known. We have included the upper Rio Negro and the Inírida basins on the basis of reliable accounts of rubber hunters and the known occurrence of the species in the Amazonas Federal Territory of neighboring southern Venezuela. However, documentation exists for the Vaupés River and Apaporis River basins based upon preserved specimens from the middle Apaporis near La Providencia (G. E. Erikson, personal collection; Universidad Nacional de Colombia, Bogotá); from 10 specimens collected by the Olalla brothers on July 9, 1928, at "Tahuapunto" (Tauapunto), on the Brazilian bank of the Vaupés River at the border of Colombia (American Museum of Natural History); and from preserved *Cacajao* mandi-

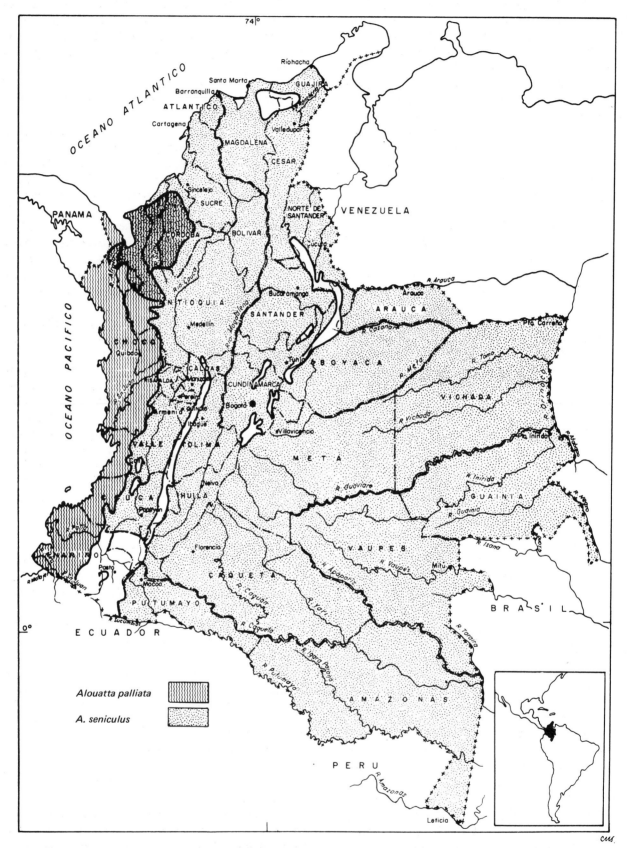

FIGURE 9   Distribution map of the genus *Alouatta* in Colombia.

bles from contemporary Indian middens in the Mitú region of the Vaupés River. Dr. Frederico Medem (*in litt.*) has reported the occurrence of this species in the Cerro de las Pinturas and El Dorado Lake regions of the Vaupés Comisaría. The western extent of the probable range of *C. melanocephalus* in Colombia is documented by a specimen collected by Mr. Carlos A. Velásquez in 1942 during the Gilliard–Dillon expedition of the American Museum (the specimen could not be located at the museum in August 1972) to the northern Macarena Mountains. In addition, Agapito, a famous chief of the Tinigua Indians, recounted in 1959 that he had killed a specimen many years earlier near La Angostura, on the southern bank of the Guayabero River, close to the southern tip of the Macarena Mountains. Accounts of a number of reliable informants who are familiar with this species in the little-explored upper Vaupés basin are in agreement that it does not occur south of the Apaporis River.

It appears likely that two subspecies of *C. melanocephalus* may exist, based on the plate published with Humboldt's original description (suggesting that the lower back and flanks are uniform in color with the hind limbs) and the description of Elliot (stating the back and sides to be reddish with a black admixture, as in the hind limbs). Specimens from the Maturaca River, a tributary of the Rio Negro, in northwestern Brazil near Venezuela, and from the Amazon Federal Territory of Venezuela, have a bright tawny back and flank color. On the other hand, all the Colombian specimens that have been examined, as well as a series from Tauapunto (Vaupés River), Brazil, and two specimens from the market in Leticia [probably originating in the area between the lower Caquetá River (or Yapura) and the lower Rio Negro on the northern side of the Amazon], are strikingly different in the markedly yellowish coloration of the sides and back without any obvious reddish tinge. If this distinction is valid, it appears that the correct subspecific designation of the Colombian form would be *C. m. ouakary* (Spix, 1823). This subspecies would agree basically with the plate of Elliot (1913), although not with his verbal description.

HABITS: It is reported by Frederico Medem, who collected the specimens from La Providencia in 1951, that *C. melanocephalus* is a high-canopy dweller, very shy, and found in groups of as many as 30 individuals. This account agrees in all respects with those of other informants and is consistent with the relatively large numbers of museum specimens collected in single localities on the same date. It is reportedly a highly prized food item by Indians of the region (e.g., Puinaves), possibly contributing to its apparent scarcity. Its food habits are essentially unknown.

STATUS: The status of *C. melanocephalus* in Colombia must be considered very precarious. Its habits and apparently savory meat do not favor its coexistence with humans in areas of settlement and degraded forest.

16. *Alouatta palliata* (J. E. Gray, 1849)—Mantled howlers.

COMMON NAMES: "Mono Zambo" in northern Caribbean coastal Colombia; "Mono Negro" in the same area as well as along the Pacific coast (sometimes also applied to *Ateles paniscus fusciceps*); "Mono Chongo" and "Chongón" in the lower Pacific coastal area adjacent to Ecuador.

DISTRIBUTION: (Figure 9) The range of *A. palliata* in Colombia includes the entire Pacific coastal lowlands. It is absent from the coastal mangrove zone and probably from the adjacent natal zone (a swampy forest type dominated by "nato," *Dimorphandra oleifera*). It inhabits most of the morphologic classes of nonflooding forest, and its upper altitudinal limit in the piedmont zone is not known. In the northern junction between the Pacific and Atlantic oceans, it occurs in the Atrato River basin and in the Sinú Valley, where it is sympatric with *A. seniculus*. A specimen from Turbaco collected by Carriker in the early 1900's and reports by Dugand of the presence of the species in Los Pendales as late as the early 1950's (both localities in the Cartagena region) suggest that its northeastern extension may even today be somewhat greater than we have chosen to indicate in our figure. However, recent field studies in the María Mountains in the departments of Sucre and northern Bolívar have not revealed any evidence of *A. palliata*, although *A. seniculus* has been observed. In the Atrato River and Sinú River basins of the Caribbean coast it occurs from pluvial through rain forests to semievergreen forests. The Colombian population is referable to *A. p. aequatorialis* (Festa, 1903).

HABITS: Very little information is available on this species in Colombia, as most areas of its occurrence are remote and have not as yet been studied.

STATUS: The status of *A. palliata* in Colombia is practically unknown, although meat hunting probably exerts more population pressure at this time than forest destruction, except in the area of its sympatry with *A. seniculus*, where intensive forest exploitation and destruction are occurring.

17. *Alouatta seniculus* (Linnaeus, 1766)—Red howlers.

COMMON NAMES: "Mono," "Mono Colorado," "Mono Cotudo," and "Cotudo" in northern and central Colombia; "Roncador" in central Colombia;

"Araguate" in Arauca, and "Bonso," "Mono Berreador," or "Berriador" in Tolima and the eastern plains; "Bongo" and "Araguato" elsewhere in the eastern plains; "Mono," "Mono Cotudo," and "Cotumono" in the Amazon region; "Guariba Vermelho" occasionally in the Leticia area (of Brazilian origin).

DISTRIBUTION: (Figure 9) *A. seniculus* is absent on the Pacific coast of Colombia and the desert of the Guajira Peninsula and has not yet been reported from the Department of Nariño (southwestern Andes near Ecuador). Otherwise it is present throughout the country, except in nonforested areas and in mountainous regions above the cloud forest belt. Its upper altitudinal limits are known to be as high as 3,200 m locally in the central Andes. It is poorly known in the Atrato basin, but specimens have been collected on the west bank of the Atrato River near Unguía as well as on the east bank near Suatatá. Its southern limits in this region are not known. Its habitat occupation is extremely varied and includes the mangrove swamps of the Caribbean coast, the gallery forests of the eastern plains and other relatively dry regions, deciduous tropical forests, rain forests, cloud forests (including oak forests), and extremely small, isolated forest patches and second growth.

Currently, the Colombian populations of *A. seniculus* are referred to the nominotypical subspecies of *A. s. seniculus* (Linnaeus, 1766).

HABITS: The group size of this highly adaptive species ranges from 2 or 3 to 15 individuals with an apparent average of 6 to 8. The species has several remarkable and obvious behavior patterns. Groups can be heard at distances of 2 km or more giving their unique vocalizations from near dawn until about 8 a.m. on clear days, and they often begin again at 4:30 or 5 p.m., prior to retiring at dusk. On cloudy days vocalizations may be heard at nearly any hour, and even on sunny days their calls may accompany the approach of a storm. Their basking behavior in exposed areas of the highest-available canopy (very often in defoliated *Ceiba pentandra* or other similarly emergent species) is also very striking. Considering their coat color (predominantly dark reddish) and the high ambient temperatures reached in nonshaded areas during midday, one wonders what body temperatures they might attain and what special cooling mechanisms they may possess beyond an extreme paucity of ventral hair. Their heavily pigmented skin undoubtedly protects them from ultraviolet radiation but should increase their absorption of heat while basking. When red howlers are disturbed, it is their habit to seek a high and concealed location in the canopy, where they remain immobile and very quiet. As has been reported

in *Alouatta palliata*, red howlers will also "rain" excrement on observers below, particularly in areas where they are not actively persecuted.

The diet of *A. seniculus* is varied in terms of plant species eaten, but generally consists of leaves and some soft fruits. Some of the principal dietary species include *Anacardium excelsum*, *Cecropia* spp., *Cassia moschata*, *Ficus* spp., and *Spondias mombin*. Stomach contents are very finely masticated, making identification of food plant species more difficult than in most other Colombian primates. *A. seniculus* is not known to have interspecific associations with other primates in Colombia with the exception of *Ateles paniscus belzebuth* (Klein and Klein, 1973). In addition, association between *A. seniculus* and *A. palliata* in the area of their sympatry is not known to us.

Red howlers are able swimmers and have been known to cross bodies of water of 200 to 300 m in width. This ability may contribute to their success in the Cauca–Magdalena delta region and other such sparsely forested and seasonally flooded areas. The same adaptive significance may be seen in their ability and willingness to cross large expanses of treeless savanna on foot. In many areas of northern Colombia and the middle Magdalena Valley, *A. seniculus* is parasitized by botfly larvae of the genus *Cuterebra*. The preferred site is the ventral neck region, and large numbers of subcutaneous *Cuterebra* pupae can be observed at times in adult *Alouatta*. While filarid nematodes are frequent findings in the peritoneal cavity of red howlers from the north coastal and eastern plains regions of Colombia, they have not as yet been observed in howler populations from the middle Magdalena Valley. With regard to reproduction, *A. seniculus* is not a particularly seasonal breeder, as dependent infants have been observed in every month from January to August. Recently (May 1972), in southern Bolívar, collections included a female about 2 months pregnant and another carrying a 2-week-old baby in localities separated by only about 80 km.

STATUS: It is difficult to generalize about the status of *A. seniculus* in Colombia. In most of the remaining forested lowlands, its call can still be heard daily. Forest destruction seems to be its principal enemy; but, due to its previously noted adaptive abilities, its continued presence is evident long after most other primate species have disappeared. Hunting of red howlers for meat is geographically variable, e.g., it is very seldom eaten in northern Colombia, but is occasionally eaten in the eastern plains by Indians and some immigrants from the mountainous interior. In the upper and middle Magdalena and Cauca valleys, it has long been prized as a food item by some segments of the human population, and today *Alouatta* is difficult to find in many suitable relic forests of this region. In these same interior areas, its hyoid bone is often used as a drinking device by campesinos who believe it to possess therapeutic properties for curing goiters. The Spanish word for goiter is "coto," and from it is derived the regionally common name "cotudo" or "mono cotudo." In the Colombian Amazon the population status of *Alouatta* is more difficult to assess, since it seems not to be common around human settlements. In March 1972, during a 2-week survey of primary rain forest areas northwest from Leticia, it was never heard or seen.

In regions of the country where the genera are both sympatric and hunted for food, *Lagothrix* and *Ateles* meat is always preferred to that of *Alouatta*. Howlers are not the basis of any commercial trade known to us in Colombia. Occasionally, live infant or juvenile specimens, produced by the hunting of adults, are reluctantly purchased by animal dealers, always at minuscule prices. Infrequently, *A. seniculus* skins are sold on a local basis, but hide hunting of this species is not presently common in Colombia.

18. *Cebus capucinus* (Linnaeus, 1758)—White-throated capuchins.

COMMON NAMES: "Micro Negro" in the departments of Córdoba, Sucre, and northern Bolívar; "Maicero Cariblanco," "Carita blanca," or "Mico" in the departments of northwestern Antioquia (Urabá region), Córdoba, Sucre, northern Bolívar, and Atlántico; "Machín" in the departments of Sucre and northern Bolívar; "Mico Maicero" in the Pacific coastal region.

DISTRIBUTION: (Figure 10) *Cebus capucinus* occurs in Colombia from the Panama border south along the Pacific coast and western slope of the western Andes; Gorgona Island off the Pacific coast (Department of Nariño); the upper Cauca Valley (very reduced populations today); northwestern Department of Antioquia (Urabá region), and the departments of Córdoba, Sucre, northern Bolívar, and southwestern Atlántico eastward to the west bank of the lower Magdalena and the middle and lower San Jorge rivers. On the Pacific coast in the Department of Valle, *C. capucinus* is absent from the coastal alluvial plain (i.e., mangrove, natal, and lowland mixed forests), but appears in the piedmont. On Gorgona Island and along the Atlantic coast, the species occurs at sea level but is not known to inhabit mangrove forest. On the western slope of the western Andes, *C. capucinus* is found at elevations as high as 1,800 to 2,100 m; it is also likely that remaining populations on the eastern slope in the Cauca Valley may attain similar altitudes. Formerly, this latter population probably extended throughout

the length of the western bank of the middle Cauca River to meet the populations of the upper San Jorge basin. Although *C. capucinus* is not known in the floodable (brackish) natal forests of the Pacific coast, it does occur in floodable (freshwater) forests elsewhere, e.g., the Atrato River basin. The species seems to prefer primary or advanced secondary forest, but it is also found in highly degraded remnant forests, including areas grazed by cattle in which some large trees and palms (particularly *Scheelea magdalenica*) survive. Other than those already mentioned, the only forest types within its range in which *C. capucinus* is not known to subsist solely are those of xerophytic character.

The taxonomic status of *C. capucinus* in Colombia suffers from lack of data in certain critical areas; however, three subspecies have been recognized, *C. c. capucinus* (Linnaeus, 1758), *C. c. curtus* (Bangs, 1905), and *C. c. nigripectus* (Elliot, 1909). The type locality of *C. c. capucinus* was fixed as northern Colombia by Goldman, based upon *Simia hypoleuca* (Humboldt, 1812), named from captive (nonpreserved) specimens seen at Puerto del Zapote northward from San Antero near the mouth of the Sinú River in the Department of Córdoba. It is likely, however, that the actual type locality was in the vicinity of Cartagena, because the species existed there during early colonial times (and was described in pre-Linnean literature), when Cartagena was the principal seaport within the range of the species. Although Hershkovitz (1949) first regarded *Simia hypoleuca* as a subspecies of *Cebus albifrons,* he later amended this interpretation (1955) by proposing "that the name be disposed of as an unavailable synonym of *C. albifrons*." Cabrera (1957), however, recognized *C. albifrons hypoleuca* as a subspecies with synonomy, including *C. malitiosus* (*C. a. malitiosus*). A close examination of Humboldt's description of *Simia hypoleuca* conclusively shows it to be conspecific with *C. capucinus* populations with an overall brownish cast such as those of the upper San Jorge Valley. Also *C. capucinus* is known from the Sinú Valley (region of Montería), as well as from the Tolú region not far from Puerto del Zapote in the coastal lowlands of the Gulf of Morrosquillo. As we will discuss later, such specimens can also be confused with particularly dark populations of *C. albifrons* from the middle Magdalena Valley.

C. c. nigripectus was described from Las Pavas (1,400 m) on the western slope of the western Andes in the Department of Valle. Later, J. A. Allen (1916) referred additional specimens from the upper Cauca Valley to this subspecies. *C. c. curtus* is based on an adult pair collected on Gorgona Island. A comparison of specimens from northern and western Colombia,

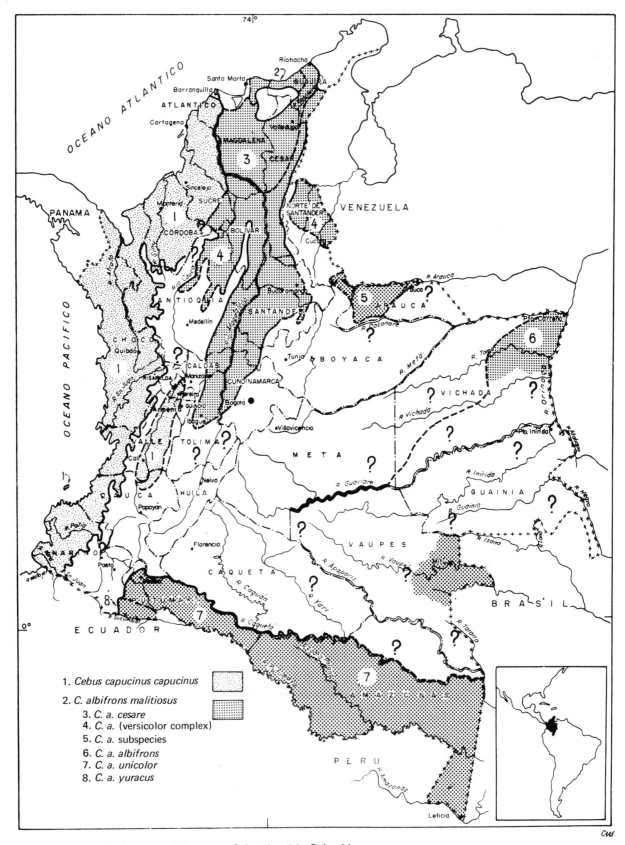

1. *Cebus capucinus capucinus*
2. *C. albifrons malitiosus*
    3. *C. a. cesare*
    4. *C. a.* (versicolor complex)
    5. *C. a.* subspecies
    6. *C. a. albifrons*
    7. *C. a. unicolor*
    8. *C. a. yuracus*

FIGURE 10   Distribution map of the genus *Cebus* (part) in Colombia.

northwestern Ecuador, and eastern and central Panama with those of Gorgona Island shows no significant size differences in adults. Differences are demonstrable in color of the light parts (from nearly pure white to white strongly suffused with tawny ochraceous), the color of the chest and belly (black to black or dark drab strongly grizzled with light-yellowish white), the extension of a yellowish latero-ventral stripe to the genital region and/or medial thighs (as sparse yellowish hairs), the degree of development of yellow in the basal portion of the ventral tail hairs, and the development and coloration of the frontal tuft in females. However, these characters are so individually variable as to make subspecific designation impossible based upon presently available material. Even the Central American populations examined by Hernández-Camacho to date are subject to this same limitation.

HABITS: The group size observed has been from about 6 to 15 individuals. As previously noted, they are found in a variety of forest conditions and can be seen from the highest canopy to very low shrublike trees and even on the ground. In this latter regard, they have been known to cross treeless expanses of several hundred meters to forage, for example, in cornfields. In degraded areas, it appears that palms such as *Scheelea magdalenica* are important in providing both food and shelter. Interspecific association of *C. capucinus* with other primates is not known in Colombia. As with other capuchins, they are omnivorous and are known to eat insects, eggs, small vertebrates, buds, leaves, berries, fruits, and (on Gorgona Island at low tide) even marine oysters, which are opened with the aid of stones (Dampier, 1697, quoted by Hill, 1960), in accounts agreeing with unpublished observations of C. R. Carpenter in Panama (personal communication, 1972). Specific seasons of reproduction for *C. capucinus* are not known.

STATUS: The species has long been exploited locally and exported as pets in small numbers in northern Colombia. It is also killed as an agricultural pest in corn-growing areas. However, as previously noted, *C. capucinus* is an adaptable primate, surviving in reduced numbers even in badly degraded habitat. Its greatest threat, as demonstrated in the Cauca Valley and the northern Department of Bolívar, is the complete destruction of forests.

19. *Cebus albifrons* (Humboldt, 1812)—White-fronted capuchins.

COMMON NAMES: "Mico," "Macaco," and "Cairara" in the Leticia region; "Mico Tangue" in the Caquetá River basin; "Maicero," "Maicero Cariblanco," and "Mico Cariblanco" throughout its non-Amazonian range in Colombia; "Carita Blanca" and "Mico Bayo" in the departments of Magdalena, Cesar, and southeastern Bolívar; "Machin" in the middle Magdalena Valley; "Ouavapavi" by the Maipures Indians, according to Humboldt.

DISTRIBUTION: (Figure 10) *Cebus albifrons* has a discontinuous (and in many areas unstudied) distribution in Colombia. It occurs in the departments of Magdalena, Cesar, southern Guajira, southeastern Bolívar (including Mompós Island), eastern Sucre between the San Jorge and lower Cauca rivers, Antioquia in the lowlands east of the Cauca River, the eastern lowlands of the Magdalena River basin, the lowlands of Santander, western Boyacá, eastern Caldas, western Cundinamarca and northern Tolima, the Catatumbo basin in North Santander Department, and the piedmont forests of the Comisaría of Arauca. In the eastern plains, it is known only from the eastern Comisaría of Vichada (the Bita River near Puerto Carreño, the lower Tomo River, and Maipures, a former mission about 3 km southeastward from the confluence of the Tuparro and Orinoco rivers just above the Maipures rapids on the Colombian bank of the Orinoco). It is absent from the eastern Comisaría of Arauca, the Casanare plains of eastern Boyacá, the Department of Meta, and the northwestern Comisaría of Vichada.

It is possible that the range of *C. albifrons* extends to the southern Comisaría of Vichada and to the comisarías of Guainia and Vaupés, since it has been reported from the southern bank of the lower Guayabero River by local informants (L. Klein, personal communication, 1972) as well as to the forests of the northern bank of the Guaviare River in the Department of Meta. Yet, it is absent from the Ariari and Guayabero basins in Meta. It may also be present in the southeastern Intendencia of Caquetá and the northern Comisaría of Amazonas and is known in all the region south of the Caquetá River. Its altitudinal range is from sea level on the Caribbean coast to elevations of 1,500–2,000 m in the Department of Tolima. For detailed descriptions of the following subspecies, the reader is referred to Hershkovitz (1949).

C. a. malitiosus (Elliot, 1909) is a well-defined subspecies inhabiting the deciduous and humid forests of the northern slopes of the Santa Marta mountains at least as high as 1,300 m. The eastern and southern limits of this population are not well defined. The subspecies is characterized largely by its rather dark-brownish overall coloration and rather light-yellowish shoulders.

C. a. cesarae (Hershkovitz, 1949) is a very light-colored and well-defined subspecies occurring in the Department of Magdalena southward from the Ciénaga Grande (it has been collected in brackish water mangroves, *Conocarpus erecta* and *Laguncularia racemosa,* on the Palenque River) and in the lowlands of the Department of Cesar from the vicinity of El Banco and Tamalameque northward to the deciduous and gallery forests of the Ranchería River, in the southern Guajira Department.

C. a. versicolor (Pucheran, 1845) is a "complex," including dark-colored populations extending along the Perijá Mountains in the southern Department of Guajira and in the Department of Cesar and occurring in the middle Magdalena Valley, including Mompós Island; in the southeastern Department of Sucre between the San Jorge and lower Cauca rivers; in the northern Department of Antioquia eastward from the Cauca River, including the Nechí River Valley; and in the humid lowlands of the Catatumbo basin in the Department of North Santander. Hershkovitz (1949) determined that three subspecies occupied this complete range: *C. a. leucocephalus* (Gray, 1865) from the eastern bank of the Magdalena River and in the Catatumbo region, characterized by a dark brown color and reduced reddish hues in the lower limbs; *C. a. pleei* (Hershkovitz, 1949) from the Norosí region in the southwest Department of Bolívar, characterized by reddishness, particularly in the limbs; and *C. a. versicolor,* recorded from the Department of Tolima and reminiscent of *C. a. pleei* but with overall light red coloring. However, in 1957 collections in the region of Barrancabermeja, on the eastern bank of the middle Magdalena River in the Department of Santander, demonstrated the presence of typical *C. a. pleei* and *C. a. leucocephalus* in localities only 10 km apart. Subsequent collections and field observations in the same region in 1958 revealed individuals typical of both supposed subspecies in the same groups as well as intermediate specimens approaching *C. a. versicolor.* As in many other populations of *C. albifrons* subspecies, a range of variation in shades of color from rather dull ("dark phase") to somewhat brighter ("light phase") individuals exists. This evidence strongly suggests that the "dark phase" (*C. a. leucocephalus*) and "light phase" (*C. a. pleei*) are extremes of the intermediate *C. a. versicolor.* This possibility must be further evaluated with specimens collected from several critical areas, particularly the western bank of the middle Magdalena River and the Lake Maracaibo drainage of Colombia.

There is a pale and dull-colored population of *Cebus albifrons* in the piedmont forests of western Arauca Comisaría, the northern tip of Boyacá, and the eastern tip of North Santander. The available specimens from the Colombian bank of the Arauca River may represent "light phase" animals of *C. a. adustus,* known from the Lake Maracaibo region of Venezuela. This possibility suggests the presence of *C. a. adustus* in the upper Apure basin of Venezuela.

C. a. albifrons (Humboldt, 1812) does not have a preserved type specimen. Judging from the original description, it was duller and less yellowish ("dark phase") than specimens recently collected in eastern Vichada Comisaría near the type locality. The type specimen, according to Humboldt, had a dark-colored tail tip, unusual for *C. albifrons* and not characteristic of any recent specimens. The present subspecies is very light-colored with a strikingly yellow tone, resembling the population from Arauca but yellower, with an orange-yellow (not brown) dorsal surface on hands and feet.

C. a. unicolor (Spix, 1823) is also a light-colored subsubspecies with a yellowish cast. It occurs in Amazonian Colombia south of the Caquetá River, except for the area of southwestern Putumayo Comisaría south of the Guamués River, which is occupied by a light, dull-brownish population referable to *C. a. yuracus* (Hershkovitz, 1949). There is a great scarcity of Amazonian study specimens, and at this time the possibility exists that *C. a. unicolor* could be a synonym of *C. a. albifrons.* Selected specimens from the Amazonas Federal Territory of Venezuela are virtually indistinguishable from *C. a. albifrons,* but the population has been referred to *C. a. unicolor* by Hershkovitz (1949).

During the past 2 years, several extremely dark-brown specimens of *Cebus albifrons* from the Barranquilla market with reported provenience in the middle San Jorge Valley have been preserved. At first inspection it is difficult to tell whether such specimens are referable to *C. capucinus* or *C. albifrons.* Intermediate characters include higher or more-recessed dark skull cap, more distinct baldness of the associated white forehead area, noticeably lighter shoulders and upper arms (all features suggesting the predominant pattern of *C. capucinus*), and an overall darker and uniformly colored body, more so than even the extremely dark individuals of *C. a. malitiosus.* By the same token, some *C. a. versicolor* (*pleei*-type) seen in the Magangué market and probably captured in the lower Cauca River region also show the above similar tendencies, except there is no noticeable increase in overall dark pigmentation. Based on these observations, as well as the several dark-brown *C. capucinus,* "intermediate-like" specimens of unknown northern Colombian provenience, it seems quite likely to us that further studies in the San Jorge Valley and other zones at the interface of *C. capucinus* and *C. albifrons* distribution may ultimately show these forms to be conspecific. Another zone critical to such an analysis is the area in northwestern Ecuador in which *C. a.*

equatorialis (J. A. Allen, 1914) and *C. capucinus* are both known to occur, but where sympatry or intergradation have yet to be determined.

HABITS: In general, the habitat requirements of *C. albifrons* do not differ from those of *C. capucinus*. The same can be said for most other species characteristics of which we are aware, with the possible exception of maximum group size. It is our impression that groups of 20–30 *C. albifrons* are common in suitable habitat. In August 1957 Hernández-Camacho observed a congregation of at least several hundred *C. albifrons* actively foraging on the leguminous fruits of *Inga* sp. at about 5:30 p.m. within a distance of about 300 m on both sides of the recently constructed road from El Centro to Quebrada Lizama in the municipality of Barrancabermeja. In this region another seemingly preferred foraging area is "guamera," a secondary forest about 15 years old with an 8–12 m canopy dominated by "guamo," also *Inga* sp. Additional important food items include the "guayabo de pava" (*Bellucia axinanthera*), which fruits throughout the year, and the continuously available food-bearing "palmichales," large homogeneous associations of palmiche (*Euterpe* sp.) common in floodable areas and boggy soils.

STATUS: In the middle Magdalena Valley and the Amazon region, *C. albifrons* is hunted for food. Throughout their range they are persecuted in areas where they forage for corn. Some commercialization of the species occurs in the Amazon region and in the north coastal Cauca River and Magdalena River delta regions. However, *C. albifrons* is a rather cautious and adaptable species. It survives in even very reduced and degraded habitat and does not make its presence obvious by conspicuous vocal or visual displays. The greatest danger to the survival of many local populations is the complete destruction of habitat for purposes of cattle-grazing and farming.

20. *Cebus apella* (Linnaeus, 1758)—Black-capped or tufted capuchins.

COMMON NAMES: "Maicero," or "Mico Maicero," is the most widespread name in Colombia; for tufted adults, "Mico" or "Maicero Cornudo," "Cachón," or "Cachudo"; in Leticia, "Mico" or "Macaco Prego"; "Aggué" (Tukano Indians, *fide* Olalla from specimen tickets).

DISTRIBUTION: (Figure 11) *Cebus apella* occurs throughout the Colombian Amazon and in the entire lowlands and piedmont (to at least 1,300 m) of eastern Colombia. It also occurs in the upper Magdalena Valley in the Department of Huila to an elevation of 2,700 m (in the region of San Agustín) and in the region of Tierradentro in the Department of Cauca at altitudes

of up to 2,500 m near Inzá. A specimen in the British Museum labeled "Tolima Mountains" and collected prior to 1900 by White may have been collected in the Huila Department, as at that time the "State" of Tolima also included Huila (e.g., specimens of *Lagothrix lagotricha lugens* collected by White and labeled "Tolima" included a latitudinal designation, 2° 20′ N, which places them in the present-day Department of Huila). It is possible that *C. apella* and *C. albifrons* may have sympatric marginal contact in the Tolima Department. In the eastern plains, *C. apella* is found in virtually every type of humid forest from gallery to palm to rain forest and in both seasonally flooded and nonflooding habitats. They are also found in broken or isolated forest, in second growth, and in mountain areas (in cloud forest). They also cross open ground in passing from one forest segment to another.

Subspecific identities of Colombian populations of *C. apella* are yet to be determined. If one follows closely the work of Hill (1960), it is necessary to accept a number of sympatric subspecies. Apparent taxonomic problems may result from well-known sexual, age–class, and individual variations in external features characteristic of this species. Until a thorough review of the situation can be accomplished, we consider it best to accept the earliest valid subspecies for northern South America, *C. a. apella* (Linnaeus, 1766), to represent all Colombian forms. We remain unconvinced, on the basis of examination of over 120 widely distributed museum specimens, that *Cebus apella* north of the Amazon from Colombia eastward exhibit phenotypic distinctions that would justify the recognition of more than one subspecies throughout this large region.

HABITS: Group size is variable, but 15–20 (at times 30) individuals seem about average for continuous forest. In isolated forest or second growth, 3–5 adults may form a group. *Cebus apella* is well known for its interspecific association with *Saimiri sciureus* (Thorington, 1967; Baldwin and Baldwin, 1971; Klein and Klein, 1973), and association with *Ateles paniscus belzebuth* has been observed in forests at the southern end of the Macarena Mountains (Hernández-Camacho, unpublished data, 1959; Klein, personal communication, 1972). They are very noisy and, except at midday, can be easily sighted or heard in undisturbed lowland forest. In areas where they have not been hunted or otherwise molested, they exhibit considerable curiosity, often approaching to within a few meters of an observer. In addition, some adults have been seen by Hernández-Camacho to throw feces and urine down toward an observer. When they travel in mature forest they seem to prefer a high route beneath the upper canopy. It appears that most birds leave an

FIGURE 11   Distribution map of *Cebus apella* in Colombia.

area when a party of *C. apella* passes through. They have been observed by Hernández-Camacho to actively grasp at species such as *Ara macao* and *Harpagus bidentatus*. On the other hand, they exhibit great fear, as evidenced by noisy flight, of such eagles as *Harpia harpyja* and *Spizaetus ornatus*. On one occasion a tayra (*Eira barbara*) was collected while in rapid pursuit of a troop of *C. apella* through the forest canopy. All large groups seem to be dominated by a large older male with younger males and juveniles occupying peripheral positions around the more central females and infants (i.e., an "age-graded male troop," Eisenberg *et al.*, 1972).

Small infants have been observed at almost every month of the year, suggesting that the species is not a seasonal breeder. In terms of food habits, they seem to eat largely fruits, including those of several palms (e.g., *Euterpe* sp., *Jessenia polycarpa*, *Scheelea* sp., and *Syagrus inajai*), of wild figs (*Ficus* spp.), of Araceae (*Monstera* sp.), wild plums (*Spondias mombin*), and berries of *Isertia* sp. (a shrub common in second growth) and of *Pourouma* sp. On occasion they have been observed to collect fallen fruits from the ground. They are also eager eaters of small Coleoptera and grubs, and it seems that they would eat any small vertebrate that they could catch. They are known for their agricultural thievery, but prefer immature corn to other fare, such as cultivated fruits.

STATUS: The greatest pressures result from habitat destruction and from active hunting of *C. apella* as agricultural pests and in some areas as a source of meat. *C. apella* is a hardy and adaptive species with an extremely wide range in Colombia. Its numbers are being reduced in the upper Magdalena Valley and the major areas of immigration and settlement in the eastern plains. It is not a highly commercialized species in Colombia, the Amazon region providing the few specimens that are exported.

21. *Lagothrix lagotricha* (Humboldt, 1812)— Humboldt's woolly monkey.

COMMON NAMES: "Mono Choyo," "Choyo," and "Choro" in the northern piedmont (Comisaría of Arauca and departments of Boyacá and Meta); "Churuco" or "Chuluco" in the upper Magdalena Valley and the comisarías of Caquetá, Vaupés, and Amazonas; "Barrigudo" in the Amazon.

DISTRIBUTION: (Figure 12) Two apparently isolated populations, marking the northernmost extension of the genus, have recently been discovered in the upper San Jorge Valley in the Department of Córdoba, around San Pedro, and in the Serranía de San Lucas, in southeastern Bolívar (see Green, 1976). To date only juvenile specimens have been examined, and the

subspecific allocation is difficult. These specimens have been very black in color with remarkably long and dense hair reminiscent of mountain populations of *L. l. lugens*.

The known populations referable to *L. l. lugens* (Elliot, 1907) are confined almost entirely to Colombia. The exception is a recently discovered population in the area of dense piedmont forest in the Sarare River drainage of the State of Apure, adjacent to the Colombian border, in Venezuela. From this location, the distribution of *L. l. lugens* continues south and west to the eastern slopes of the Andes, including the piedmont and neighboring wide gallery forests. Its range extends easterly in the plains along the Ariari River to include the Macarena Mountains. It is replaced by *L. l. lagotricha* on the northern bank of the lower Guayabero River. The subspecies of *L. l. lugens* extends westerly across the eastern Andes to include the upper Magdalena Valley, reaching an upper altitudinal limit of about 3,000 m. It is known as far north as the upper Saldaña Valley on the west side of the Magdalena Valley in southern Tolima Department. On the east side of the Magdalena River, it occurs as far north as the mountains of the Pandi area, in southern Cundinamarca Department. As indicated in the distribution map for this species, its existence is open to question along the eastern slope of the central Andes, which is a possible historical or present corridor to the recently discovered northernmost populations in southeastern Bolívar. The southernmost known occurrence of *L. l. lugens* is in the area of Mocoa, in the piedmont forest of northwestern Putumayo Comisaría. As the adjacent southern piedmont region of the Putumayo basin and the neighboring Aguarico River drainage of Ecuador provides similar habitat and is poorly known, it is possible that *L. l. lugens* will yet be found in Ecuador.

The populations of *L. l. lagotricha* in Colombia are continuous with those of northwestern Brazil, extreme northern Peru, and extreme northeastern Ecuador. Its northernmost extension in Colombia is at least to the northern bank of the Guaviare River of the Orinoco drainage. Its northwestern extension meets the southeastern extension of *L. l. lugens* approximately at the junction of the piedmont forest with lowland rain forest. Questions of subspecific sympatry are problematic, as observations in the region of Florencia and the Orteguaza River basin in the Intendencia of Caquetá suggest a gradual rather than an abrupt intergradation (represented by a changing frequency of buffy brown hair coats characteristic of typical *L. l. lagotricha* with the more grayish-to-blackish coat characteristic of typical *L. l. lugens*). Some individuals appear approximately intermediate, but specimens

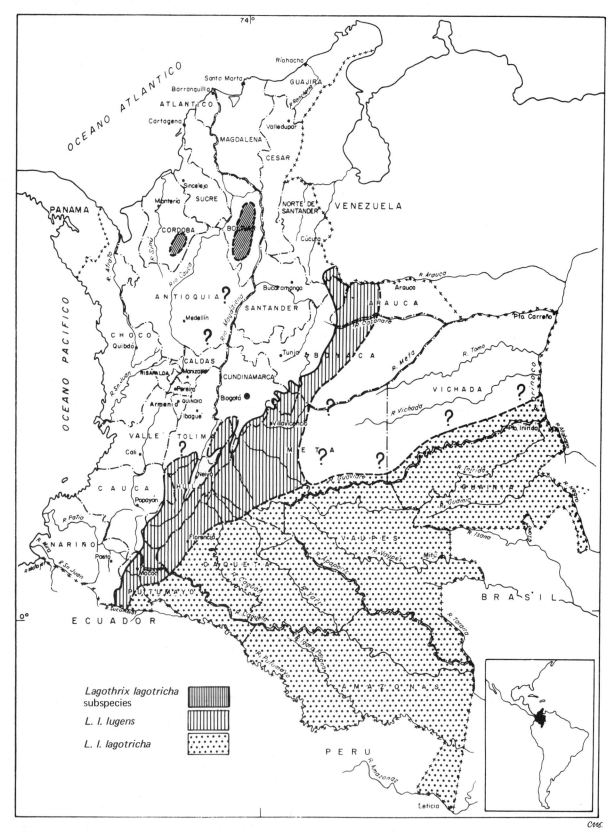

FIGURE 12 Distribution map of the genus *Lagothrix* in Colombia.

showing all characteristics can be found in the same locality, if not in the same groups.

The habitat of this species is always some type of humid forest, e.g., the gallery and palm (*Mauritia flexuosa* association) forests of the eastern plains, rain forest (seasonally flooded or not), and cloud forest. It can be observed in undisturbed primary forest or somewhat degraded forest, but we have never seen this species in second growth or young secondary forest.

HABITS: Woolly monkeys have been seen in groups of four to six individuals (which have always included a rather old adult male) in piedmont forests of eastern Colombia and in the upper Magdalena Valley. They seem to be seen most frequently just beneath the upper canopy and sometimes in emergent trees. When disturbed by observers, they tend to move away from the area as opposed to hiding, a tactic more common in *Alouatta*. They also may void excreta and tend to use their hands in throwing it toward an observer. Adult males give threatening displays including branch shaking and grunting vocalizations, often with canines exposed.

The diet of *Lagothrix*, insofar as we have observed, includes only vegetable matter but not leaves. In *L. l. lagotricha* from the eastern plains, stomach contents have always contained quantities of various palm fruits, particularly *Mauritia flexuosa*, *Attalea regia* (syn.: *Maximiliana elegans*), and *Jessenia polycarpa*. During the fruiting season of *Jessenia* (February–April) and *Attalea* (June–July), areas of these palm populations are good places to encounter *Lagothrix*. A recently collected adult male specimen of *L. l. lugens* from the upper Magdalena Valley at more than 2,000-m elevation had stomach contents that included fruits of a Clusiacae (*Clusia* sp.), *Ficus* sp., and an unidentified species (probably a Sapotaceae). The morphology of its stomach (long and cylindrical with at least three distinct mucosal regions) was much like that of *Ateles* and quite different from that of *Alouatta*.

Interspecific associations with other primates are not known in Colombia. In fact we have the distinct impression that *Lagothrix* and *Ateles* are largely mutually exclusive. They do at times occur in the same region and in apparently identical habitat but have not been observed by Hernández-Camacho closer together than distances of 5 km or more. In the Caquetá and Putumayo regions, two species of eagles are known as "Aguilas Churuqueras" (*Harpia harpyja* and *Morphnus guianensis*) due to their reputation as *Lagothrix* predators. Carlos Lehmann (1959) has also reported the predation of *Lagothrix* by Isidore's eagle (*Oroaetus isidori*) in the Moscopán region near the Puracé National Park in the central Andes. *L. l.*

lagotricha collected from the eastern plains of Colombia by Hernández-Camacho have always had an abundance of filarid nematodes in the abdominal cavity.

STATUS: In total, *L. lagotricha* is probably the most persecuted primate species in Colombia. Its meat is highly prized as food by some people from the mountainous interior of Colombia (many of whom have emigrated to the eastern plains) and particularly by Indians and mestizos in the Amazon (see Izawa, 1972). The meat of *L. l. lagotricha* is particularly esteemed at the peak of the fruiting season of the palms on which they feed because of the subcutaneous fat the monkeys accumulate (see also Hershkovitz, 1972). As *Lagothrix* is known only in primary forest, its future is also closely linked to the survival of undisturbed forests. Originally, as an adjunct to meat hunting, immature *Lagothrix* were maintained and often sold as pets.

Today it is likely that the situation is reversed in some areas of the Colombian Amazon, with meat being an adjunct to the collection of immature *Lagothrix* for sale as pets. Recently, at the airport of Barranquilla, a man from Puerto Leguízamo, on the Putumayo River, was observed with three juvenile woolly monkeys and a few other small mammals and reptiles on his way to sell them to a local animal dealer. He had started his journey from Puerto Leguízamo by plane less than 12 hours earlier and would return home with a considerable profit (even after absorbing the price of round-trip plane fare), largely due to receipts from the three young *Lagothrix*. Of the more than 300 young woolly monkeys legally exported from Colombia in 1970, the majority probably came from the Putumayo and Caquetá regions. The *Lagothrix* exported from Leticia are seldom, if ever, captured in Colombia; most come from Amazonian Brazil and Peru and represent *L. l. poeppigi* (Schinz, 1844). Recent field studies in the lower trapezium region revealed no populations of *Lagothrix* and no Colombian inhabitants of the area who knew where they might exist. That such reduced or absent populations have not always been the case is demonstrated by the report of Bates in 1863 that the Ticuna Indians of Tabatinga (a Brazilian town very near Leticia on the same bank of the Amazon) killed at least 1,200 woolly monkeys a year for food.

22. *Ateles paniscus* (Linnaeus, 1758)—Spider monkeys.

COMMON NAMES: "Marimonda" and "Marimunda" in northern, central, and eastern Colombia and in the Comisaría of Guainía; "Marimba" in the eastern plains as well as central and Amazonian

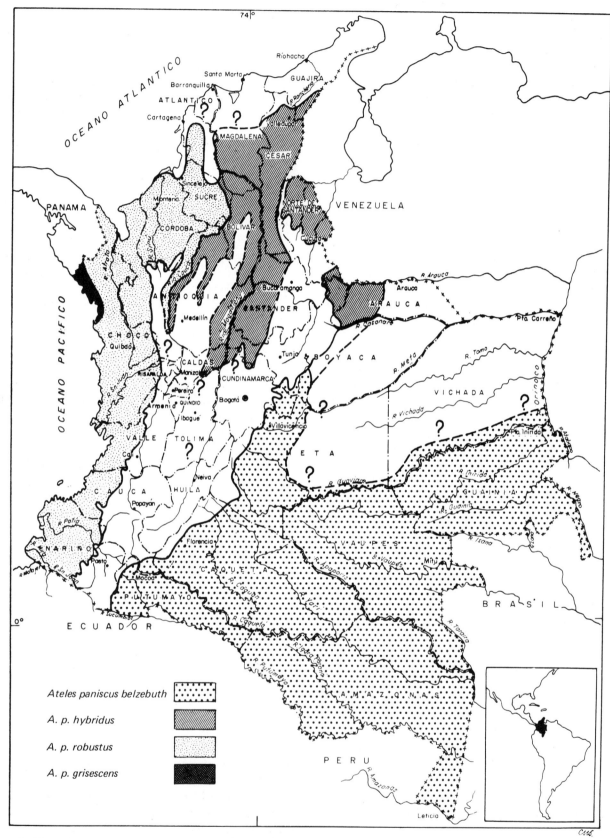

FIGURE 13   **Distribution map of the genus _Ateles_ in Colombia.**

Colombia; "Choiba" in the Department of Antioquia and the middle Magdalena Valley; "Braceadora" in the eastern plains; "Coatá" in the Leticia region (Brazilian derivation); "Maquizapa" in the Putumayo and Leticia regions (of Keshwa origin); "Mono Negro" in the Department of Chocó and on the Pacific coast; "Mica" in the Department of Bolívar; "Zamba" in the Department of Antioquia (for *A. p. robustus*).

DISTRIBUTION: (Figure 13) In Colombia, *Ateles* is found from sea level to maximum elevations of 2,000–2,500 m, the latter locally on the western slope of the western Andes and in southern Santander Department. It also occurs to 1,400 m in the Perijá Mountains in the departments of Cesar and Guajira and to 1,300 m on the eastern slope of the eastern Andes in Boyacá; otherwise, it is generally encountered at lower than 800 m. Spider monkeys are notably absent in the dryer districts of the Caribbean coast, the northern and western wet slopes of the Santa Marta Mountains, the upper Cauca and Magdalena valleys, large areas of gallery forest in the northeastern plains, and locally over considerable extensions of the Amazon basin.

A. paniscus belzebuth (Humboldt, 1812) in general has a naked face with black skin, sometimes slightly depigmented around the orbits and muzzle. It usually has a rather conspicuous submalar stripe of white hair that terminates in front of the base of the ear, as well as an often conspicuous triangular white-to-yellowish frontal hair patch, which may be infused with dark hairs (even to the point of disappearance). The upper parts, including the forelimbs and hands and most of the tail, are usually black. The tail tip is variably yellowish to black, although the undersurface is often buffy-yellowish to orangish, as are the underparts in general, usually including most of the leg. The feet are always black, as well as the hair covering the knees, which may also extend variable distances onto the anterolateral surfaces of the thigh and lower leg. A remarkable variation in many of these color characters may occur locally and individually.

The range of *A. p. belzebuth* includes the Amazon lowlands northward at least to the Guaviare River and also the Macarena Mountains and the piedmont forests (locally to 1,300 m) northward to the Upía River drainage in southern Boyacá Department. The subspecific identification of populations occurring in the piedmont north of the Upía River to the border of the Comisaría of Arauca (where *A. p. hybridus* occurs) is not yet known. In addition, *A. p. belzebuth* was recorded by Humboldt and Bonpland (1812) from both the Atures and Maipures rapids of the Orinoco River. As they did not mention on which bank of the river the species was observed or collected, it is uncertain whether this record is referable to Venezuela or to Colombia territory in the present Comisaría of Vichada.

Furthermore, Brother Apolinar María (1913) recorded a specimen from "Tolima" in the upper Magdalena Valley that he designated as *Ateles "beelzebuth."* The specimen unfortunately is no longer available, but we are somewhat assured by his report in the same paper of *A. "hybridus"* from farther north in the Magdalena Valley (Minero River region of western Boyacá Department) that the former was a distinctly darker specimen and possibly identical with *A. p. belzebuth* of the eastern piedmont of the eastern Andes. Although no other records are known to us of *A. p. belzebuth* west of the eastern plains and piedmont, a precedent exists for such "passage" into the upper Magdalena Valley in the cases of *Cebus apella*, *Lagothrix lagotricha lugens*, and *Saimiri sciureus caquetensis*, as well as several other species of arboreal and terrestrial vertebrates.

A. paniscus hybridus (I. Geoffroy-St. Hilaire, 1829) in general also has an approximately naked face as in *A. p. belzebuth* but usually does not have conspicuous depigmented areas in adults. The submalar stripe and the frontal triangle are also variably present but in general not as conspicuously so as in *A. p. belzebuth*. The color of the back ranges from light grayish-brown (or avellaneous) to rather rich brown. The head, neck, forelimbs, and upper surface of the tail are invariably darker than the back and in some specimens approach blackish-brown. The hind limbs are usually lighter and similar in color to the back, with the exception of typically darker coloration over the knees. In the darkest specimens known, referable to the population of the western bank of the Magdalena River, the entire lateral surface of the hind limb is dark brown and relatively uniform with the color of the back and feet. The underparts of *A. p. hybridus* vary from dirty-white to buffy, not contrasting markedly with the flanks in populations east of the Magdalena River. The ventral hair of the tail ranges from yellowish to buffy light brown. The contrast of the darker color of the extremities, tail, and head with the lighter back is variably evident. The eye color (iris) of *A. p. hybridus* is usually light brown but is sometimes grayish-blue. Some specimens from the Cesar Valley have an overall bleached coloration, appearing light buffy-gray. An extremely light, uniformly buffy-colored specimen with bluish eyes was also seen in captivity by Hernández-Camacho in Caño Muerto in the middle Magdalena Valley. In addition, a completely white specimen with pink skin and light blue eyes has been exhibited at the Barranquilla Zoological Garden and variously reported in popular literature by R. Tinoco, the Zoo Director.

A somewhat remarkable fact is that all known specimens west of the Magdalena River are considerably darker than those in essentially identical habitat east of the river. The range of *A. p. hybridus* includes the eastern bank of the lower Cauca River basin, the departments of Magdalena and Cesar (northward to the southernmost slopes of the Santa Marta Mountains), the northernmost extension of the Perijá Mountains in the Department of Guajira, and the middle Magdalena River region at least to the northern departments of Caldas and Cundinamarca. There are also at least two additional populations in Colombia on both flanks of the eastern Andes Mountains at the border with Venezuela; one occurs in the Catatumbo River basin of North Santander Department and the other in the northeastern piedmont in the Comisaría of Arauca.

A. paniscus robustus (J. A. Allen, 1913) also has a black face that is essentially hairless, at times with some depigmentation of the orbits and nasal orifices. The white submalar stripe is either entirely missing or represented by only a few scattered white hairs. The triangular frontal patch is usually entirely absent but apparently is present in some local populations. The hair coat is completely black, sometimes with very slight brownish shades on the head and back. Scattered yellowish hairs can often be found on the ventral surface of the body and inner surface of the thighs. One captive specimen, reputed to be from the Atrato River basin, was described as having reddish underparts and given the name of *A. "rufiventris"* (Sclater, 1872). Eye color is usually dark brown for *A. p. robustus*, but rarely may be lighter to even bluish. The distribution of the subspecies covers the entire Pacific lowlands (except the region around Juradó in northwestern Chocó Department); the Urabá region of northwestern Antioquia Department; the departments of Córdoba, Sucre, and northern Bolívar eastward to the lower Cauca River and along the western bank to south-central Antioquia (the most southerly record is from Concordia). In recent times the northernmost limit was the southern bank of the Canal del Dique in the Cartagena region; however, it probably formerly occurred northward to the Pendales region, where some luxurious hygrotropophytic forest still survives. This subspecies occupies the greatest range of habitat types of all of the Colombian forms of *Ateles*, from hygrotropophytic through pluvial to cloud forests. The southernmost specimen known in Colombia is from Barbacoas in the Department of Nariño. Interestingly, it does not show any of the brownish coloration typical of the Ecuadorian population of *A. p. fusciceps* (a form that may yet be found in Colombia in the Mira River valley of Nariño Department at the Ecuador border). Of historical interest is the fact that most old records of *A. p. robustus* are reported under the species designation of *A. ater* (F. Cuvier, 1823).

A. p. grisescens (J. E. Gray, 1866) is characterized by a brownish or rusty-colored back (with black hair tips) and by completely black head, limbs, and tail. It could well be considered a traditional form between the bright yellowish-colored *A. p. geoffroyi* group of middle America and the largely jet black *A. p. robustus* of southeastern Panama and adjacent Colombia. *A. p. grisescens* is known in Colombia only from the vicinity of Juradó very near the Panamanian border on the Pacific coast. It is undoubtedly restricted by the Baudó Mountains to a narrow coastal strip that may extend as far south as Cabo Corrientes.

From the above descriptions of Colombian *Ateles* populations, the fact that we consider all *Ateles* to be conspecific, as well as some of the logic underlying our conclusion, should be apparent. This position is also consistent with that of Hershkovitz (1969, 1970) and underscores the fact that all forms of the genus are allopatric with fundamental differences only in color. The characters of several supposedly distinct "species" recognized by Kellogg and Goldman (1944) show intergradation, e.g., *Ateles geoffroyi* intergrades towards *Ateles fusciceps* through *A. g. grisescens;* a specimen from Catival, San Jorge River, Colombia, representing *A. fusciceps robustus* shows a decided tendency toward *A. belzebuth hybridus* with a strong admixture of light-colored hairs on the back; *A. belzebuth belzebuth* has considerable individual and local variation in the development of dark brownish or black areas of the upperparts; and Bartlett (cited by Elliot, 1913) has recorded a specimen of *A. paniscus chamek* from Chamicuros in the Huallaga River of Peru, "which had the thighs and belly very gray or grizzled," thus approaching *A. belzebuth*.

HABITS: Spider monkeys are known to occur in social groups of variable size. Some recent information (Klein, 1971a; Klein and Klein, 1976) suggests that groups of 15–30 may break into subunits for much of the day and regroup later in the afternoon. In remnant or degraded forests, it is not uncommon to find smaller groups that appear to consist of at least one adult pair and associated offspring and/or a few other individuals. Activity can be observed from very early morning throughout the day to sunset. The animals are usually observed at middle or high-canopy levels, but it is not uncommon for them to forage on the ground in some circumstances. At times, when disturbed, adults are known to drop or throw twigs and small branches and to vocally threaten an observer. They are not as difficult to follow in level, open-floored forest as one might imagine, as they are exceptional neither in speed nor stealth. They tend to occur in more mature forest and are not usually found in historically isolated small forests (eastern plains) as are *Cebus, Saimiri,* and *Callicebus* at times.

On the other hand, in north-coastal Colombia *Ateles* is often encountered in recently isolated, relatively mature forest tracts. They are essentially vegetarian and can become quite fat during the fruiting season of some plants (particularly a palm called "seje," *Jessenia polycarpa*). *Ateles* also favors fruits of the genus *Ficus* and *Brosimum,* as well as *Spondias mombin,* which are often swallowed with little or no mastication. As previously noted, their stomachs are remarkably large and contain several distinct mucosal regions. In the middle Magdalena Valley, *A. p. hybridus* has been seen with newborn infants in August, while in the Macarena Mountains *A. p. belzebuth* has been observed with the same in February. It seems possible that spider monkeys are not highly seasonal breeders in Colombia, although definitive information is probably lacking (see also Klein, 1971b).

STATUS: In large areas of Colombia, meat hunting is a greater threat to *Ateles* than its habitat destruction, even though the latter pressure is often quite great. Spider monkeys seem not to be actively hunted for food in the Caribbean coastal region or by descendants of the original settlers of the eastern plains. However, they are hunted by recent settlers originating from the mountainous interior and by all known indigenous groups within their range, including those in the Colombian Amazon. In many areas of apparently suitable forest where *Ateles* is not found today, it seems possible that this absence could be related to historic hunting pressures, to certain rather subtle ecological deficiencies (see Klein and Klein, 1976), and/or to sylvatic yellow fever in recent times. Some hunting of *Ateles* for exportation does occur in the Caribbean coastal and Amazon regions, but this trade in Colombia is not very large. As with woolly monkeys, most spider monkeys in the Leticia market originate from Brazil or Peru and belong either to the *A. p. chamek* group of black spider monkeys or the *A. p. belzebuth* group with a prominent forehead triangle.

23. Species of Questionable Occurrence in Colombia
    A. *Chiropotes satanas* (Hoffmansegg, 1807)— Black saki.

This primate form may yet be confirmed as occurring in Colombia based largely on records of its existence on the bank of the Orinoco River. All pertinent records known to us are from the Venezuelan banks (in the Amazon Federal Territory of Venezuela) of the Orinoco (right bank), Casiquiare (eastern bank), and Negro (eastern bank) rivers, or it is uncertain from which bank such collections originated. If *Chiropotes* occurs in Colombia it would be on the eastern border of Guainia Comisaría or possibly the southeastern tip of Vichada Comisaría on the left banks of the Orinoco or Atabapo (Orinoco tributary) rivers or on the right bank of the upper Negro (or Guainía) River. A specimen in the American Museum from Pitado (5 km above San Fernando de Atabapo) on the Orinoco above the mouth of the Atabapo River is the closest known locality of *C. satanas chiropotes* (Humboldt, 1812) to Colombia. If the collection occurred on the left bank, the locality would be less than 10 km (without limiting geographic barriers) from Colombia.

B. Other primate forms. It is possible, although not too likely, that *Cacajao calvus* (Spix, 1823) and/or *Saguinus labiatus thomasi* (Goeldi, 1907) exist as far west as the Amazonas Comisaría in Colombia between the Caquetá and Putumayo rivers. However, the nineteenth-century records of *Saguinus mystax* (Spix, 1823), *S. bicolor* (Spix, 1823), and *Cacajao rubicundus* (I. Geoffroy-St. Hilaire, 1848) from Pebas, on the northern bank of the Amazon in the present Department of Loreto, Peru (which could imply an extension into the Amazonas Comisaría trapezium of Colombia), are incorrect. In addition, the type specimen of *S. fuscicollis illigeri* (Pucheran, 1845) was incorrectly assumed to have come from Colombia. Another record (Elliot, 1913) for *C. rubicundus* from the northern bank of the Amazon opposite Sao Paulo de Olivenca, in Brazil (somewhat northeast from Leticia, Colombia), is also in error, as this form is not known north of the Amazon River.

APPENDIX: SOME VERNACULAR PRIMATE NAMES USED BY VARIOUS INDIAN TRIBES IN COLOMBIA

Achagua Indians, Comisaría of Arauca and eastern Department of Boyacá: *A. seniculus,* "Arabata" (L. M. Girón, 1883).

Andaquí (Andakí) Indians, southern Department of Cauca and extreme western Intendencia of Caquetá: *A. p. belzebuth,* "Fiaguai" (Castellví, 1938).

Catío Indians, northwestern Department of Antioquia: *A. seniculus,* "Kuara" (Fray P. del Santísimo Sacramento, in Ortiz, 1940).

Chocó Indians, Department of Chocó: *A. trivirgatus,* "Uná"; *A. seniculus,* "Chipuro"; *A. p. robustus,* "Hierré," "Perré," "Yerré" (Vallejo, 1929).

Coreguaje Indians, Intendencia of Caquetá: *A. seniculus,* "Emo," "Emú," "Emorí," "Emuri"; *Lagothrix,* "Nasé"; *A. p. belzebuth,* "Painaso" (Castellvi, 1938).

Cubeo Indians, Mitú region, Comisaría of Vaupés: *S. inustus*(?), "Abujijiyo"; *Saimiri,* "Jijiyo"; *C. torquatus,* "Wao"; *A. seniculus,* "Emu"; *C. albifrons,* "Waja"; *C. apella,* "Taque"; *Lagothrix,* "Caparo" (Salser, 1972, Instituto Lingüístico de Verano).

Guahibo Indians, Comisaría of Vichada: *Saimiri*, "Titi" and Tséle"; *Aotus*, "Mucúali"; *C. torquatus*, "Ojó-ojo"; *A. seniculus*, "Néjë"; *C. albifrons*, "Vánali"; *C. apella*, "Pabábë"; *A. p. belzebuth*, "Cuvéri" and "Cuváiri"; *Lagothrix*, "Capálu" (Kondo, 1972, Instituto Linguístico de Verano). *Saimiri*, "Chuleyo" (Fernández and Bartolomé, 1895).

Guaké Indians: *A. seniculus*, "Arabata"; *Lagothrix*, "Arimina," "Ariminá"; *A. paniscus*, "Jarachi" (Castellví, 1938).

Huitoto Indians, Curiplaya, Caquetá Intendencia: *Lagothrix*, "Jemo." La Chorrera, Igara-Paraná River, Amazonas Comisaría: *Lagothrix*, "Xemma" (Castellví, 1938).

Karidaker Indians: *Lagothrix*, "Caparro" (Humboldt, 1812).

Maipures Indians, vanished tribe of the Colombian bank of the Orinoco River, Comisaría of Vichada: *Saimiri*, "Bitschetchies," "Bichechíes" (Humboldt, 1812).

Miraña Indians (Bora group), southeastern Comisaría of Caquetá: *A. seniculus*, "Inomé" (Castellví, 1938).

Muinane Indians* (Bora group), La Sabana and Araracuara, Amazonas Comisaría: *Cebuella*, "Jibiillaje"; *S. nigricollis* and/or *S. fuscicollis* and *Callimico*, "Júhusaryje"; *Saimiri*, "Tilli"; *Aotus*, "Tóhomimi"; *C. torquatus*, "Gaahi"; *Pithecia*, "Júubaiga"; *A. seniculus*, "iju"; *C. albifrons*, "Jimuhai"; *C. apella* "Chuyiyi"; *A. p. belzebuth*, "Méecu"; *Lagothrix*, "Ciimi" (Walton, 1972, Instituto Linguístico de Verano).

Muzo Indians, a vanished tribe of western Department of Boyacá: *Aotus*, "Kubaime" (Simón, 1882–1892).

Noanamá Indians, Department of Chocó: *A. villosa*, "Kotudú" (Ortiz, 1940).

Puinave Indians, Comisaría of Vaupés: *Saimiri*, "Chiau," "Koka"; *Aotus*, "Macuriya" (also Makú Indians); *Callicebus* sp., "Tuú"; *A. seniculus*, "Caa"; *Lagothrix*, "Choicack"; *A. p. belzebuth*, "Chairi" (Cabrera, unpublished).

Quechua (Keshwa), dialect of the Ingano Indians, Putumayo Comisaría: *Aotus*, "Tutamono" (also Kamsá, Sibundoy Indians, Putumayo Comisaría); *A. seniculus*, "Koto," "Kotu," "Kotomonu" (Castellví, 1938).

Quimbaya Indians, a vanished tribe of the departments of Quindío and Risaralda: *A. seniculus*, "Chifurrú" (Bastian, 1878).

Siona Indians, Comisaría of Putumayo: *Aotus*, "Unamihué"; *A. seniculus*, "Emú"; *Lagothrix*, "Guaó," "Mazo"; *A. p. belzebuth*, "Painnazo" (Castellví, 1938).

Tunebo Indians, Sarare River region, northern Department of Boyacá, southeastern Department of North Santander; *Lagothrix*, "Savároma"; *A. p. hybridus*, "Citroa" (Castellví, 1938).

Yucuna Indians, El Depósito, Mirití Paraná River, Amazonas Comisaría: *Saimiri*, "Cuhuijru"; *C. apella*, "Calapichi"; *Lagothrix*, "Caparu" (Schauer, 1972, Instituto Linguístico de Verano. Also Vaupés Comisaría: *Saimiri*, "Cuhisú" (Cabrera, unpublished).

Indians of uppermost San Juan River, Department of Risaralda: *A. seniculus*, "Truá" (Vallejo, 1929).

---

*h = phonetic "glottal stop" of Muinane Indian language; ë or i = phonetic high central unrounded vowel of Guahibo and Muinane languages.

Bibliography to Appendix

Bastian, A. 1878. Die Kulturlander des alten Amerikas. T. I. Berlin, p. 243, Nota 1.

Cabrera, I. Unpublished. Information based on recent field studies in Colombia.

Castellví, M. de. 1938. Materials para estudios glotologicos. Bol. Estud. Hist. 11(84):368–372. Pasto.

Fernández, M., and M. Bartolomé. 1895. Ensayo de gramatica hispanogoahiba. Bogota.

Girón, L. M. 1883. Los Achaguas, Papel Periodico Ilustrado, Bogota.

Humboldt, A. de, and A. Bonpland. 1812. Recueil d'observations de zoologie et d'anatomie comparée. Smith and Gide, Paris.

Kondo, R. 1972. Personal communication.

Ortiz, S. E. 1940. Lingüística Colombiana, familia Choko. Univ. Catol. Bolivar., 6(18):46–77. Medellin.

Salser, J. K. 1972. Personal communication.

Schauer, J. 1972. Personal communication.

Simón, Fray Pedro. 1882–1892. Noticias historiales de las conquistas de Tierra Firme en las Indias Occidentales, 5 vols. Bogota.

Vallejo-E., J. 1929. Vocabulario baudó, revista de Colombia, Bogota, p. 134. Reprod. in Idearium, vol. I., Pasto, 1937, p. 259.

Walton, J. 1972. Personal communication.

REFERENCES

Allen, J. A. 1916. Mammals collected on the Roosevelt Brazilian expedition, with field notes by Leo E. Miller. Bull. Am. Mus. Nat. Hist. 35:559–610.

Alston, E. R. 1879. Mammalia. *In* Biologia Centrali-Americana, vol. 1, with an introduction by P. L. Sclater. pp. I–XX, 1–220 tables (col.) 1–22, Taylor and Francis, London.

Baldwin, J. D., and J. I. Baldwin. 1971. Squirrel monkeys (*Saimiri*) in natural habitats in Panama, Colombia, Brazil and Peru. Primates 12(1):45–61.

Bates, H. W. 1863. The naturalist on the River Amazonas, vol. II. John Murray, London.

Cabrera, A. 1957. Catálogo de los mamiferos de America del Sur Instituto Nacional de Investigacion de la Ciencias Naturales, Ciencio Zoologica. IV:No. 1, Buenos Aires and Peru.

Cooper, R. W. 1968. Squirrel monkey taxonomy and supply. Pages 1–29 *In* L. A. Rosenblum and R. W. Cooper, eds. The squirrel monkey. Academic Press, New York.

Cooper, R. W., and J. Hernández-Camacho. 1975. A current appraisal of Colombia's primate resources. Pages 37–66 *in* G. Bermant and D. G. Lindburg, eds. Primate utilization and conservation. John Wiley and Sons, New York.

Eisenberg, J. F., N. A. Muckenhirn, and R. Rudran, 1972. The relation between ecology and social structure in primates. Science 176:863–874.

Elliot, D. G. 1913. A review of the primates, vols. I and II. American Museum of Natural History, New York.

Erikson, G. E. 1963. Brachiation in New World monkeys and in anthropoid apes. Symp. Zool. Soc. Lond. 10:135–164.

Fooden, J. 1963. A revision of the woolly monkeys (genus *Lagothrix*). J. Mammal. 44:213–247.

Fooden, J. 1964. Stomach contents and gastrointestinal proportions in wild-shot Guianan monkeys. Am. J. Phys. Anthropol. (n.s.) 22:227–231.

Green, K. M. 1976. The nonhuman primate trade in Colombia. This volume.

Hernández-Camacho, J., and E. Barriga-Bonilla. 1966. Hallazgo del genero *Callimico* (Mammalia: Primates) en Colombia. Caldasia 9(44):365–377.

Hershkovitz, P. 1949. Mammals of northern Colombia. Preliminary report no. 4: Monkeys (Primates), with taxonomic revision of some forms. Proc. U.S. Nat. Mus. 98:323–427.

Hershkovitz, P. 1955. Notes on American monkeys of the genus *Cebus*. J. Mammal. 36(3):449–452.

Hershkovitz, P. 1963. A systematic and zoogeographic account of the monkeys of the genus *Callicebus* (Cebidae) of the Amazonas and Orinoco River basins. Mammalia 27(1):1–79.

Hershkovitz, P. 1966. Taxonomic notes on tamarins; genus *Saguinus* (Callithricidae, Primates) with descriptions of four new forms. Folia Primatol. 4:381–395.

Hershkovitz, P. 1968. Metachromism or the principle of evolutionary change in mammalian tegumentary colors. Evolution 22:556–575.

Hershkovitz, P. 1969. The evolution of mammals in southern continents. VI. The recent mammals of the neotropical region: a zoogeographic and ecological review. Q. Rev. Biol. 44(1):1–70.

Hershkovitz, P. 1970. Notes on tertiary platyrrhine monkeys and description of a new genus from the late Miocene of Colombia. Folia Primatol. 12:1–37.

Hershkovitz, P. 1972. Notes on New World monkeys. Int. Zoo Yearb. 12:3–12.

Hill, W. C. O. 1960. Primates, IV. Cebidae, Part A, Edinburgh Univ. Press.

Humboldt, A. de, and A. Bonpland. 1812. Recueil d'observations de zoologie et d'anatomie comparée. Smith and Gide, Paris.

Izawa, K. 1972. Monkeys in the upper basin of Amazon. Monkey 16(2):5–18. (In Japanese)

Kellogg, R., and E. A. Goldman, 1944. Review of the spider monkeys. Proc. U.S. Nat. Mus. 96:1–45.

Klein, L. L. 1971a. Ecological correlates of social grouping in Colombian spider monkeys. Paper presented at the Animal Behavior Society Meeting, Logan, Utah.

Klein, L. L. 1971b. Observations on copulation and seasonal reproduction of two species of spider monkeys, *Ateles belzebuth* and *A. geoffroyi*. Folia Primatol. 15:233–248.

Klein, L. L., and D. J. Klein. 1973. Observations of two types of neotropical primate intertaxa associations. Am. J. Phys. Anthropol. 38(2):649–653.

Klein, L. L., and D. J. Klein. 1974. Neotropical primates: aspects of habitat usage, population density, and regional distribution in La Macarena, Colombia. This volume.

Lehmann, F. Carlos, 1959. Contribuciones al estudio de la fauna de Colombia. XIV. Neuvas observaciones sobre *Oroaetus isidori* (Des Murs). Nov. Colomb., Contrib. Cient. Mus. Hist. Nat. Univ. Cauca, Popayan 1(4):169–195.

María, Brother Appolinar. 1913. Catálogo del Museo (del Instituto de La Salle, Bogotá). Bol. Soc. Cienc. Nat. Inst. La Salle, Bogotá, 1(1).

Mason, W. A. 1966. Social organization of the South American monkey *Callicebus moloch*: a preliminary report. Tulane Stud. Zool. 13:23–28.

Mason, W. A. 1971. Field and laboratory studies of social organization in *Saimiri* and *Callicebus*. Primate Behavior 2:107–137.

Moynihan, M. 1964. Some behavior patterns of platyrrhine monkeys. I. The night monkey (*Aotus trivirgatus*). Smithson. Misc. Coll. 146(5):1–84.

Moynihan, M. 1966. Communication in the titi monkey, *Callicebus*. J. Zool. (Lond.) 150:77–127.

Moynihan, M. 1967. Comparative aspects of communication in New World Primates. *In* D. Morris, ed. Primate ethology. Weidenfield and Nicolson, London.

Moynihan, M. 1970. Some behavior patterns of platyrrhine monkeys. II. *Saguinus geoffroyi* and some other tamarins. Smithson. Contrib. Zool. 28:1–77.

Moynihan, M. 1976. Notes on the ecology and behavior of the pygmy marmoset, *Cebuella pygmaea*, in Amazonian Colombia. This volume.

Thorington, R. W., Jr. 1967. Feeding and activity of *Cebus* and *Saimiri* in a Colombian forest. Pages 180–184 *In* D. Starck, R. Schneider, and H. J. Kuhn, eds. Neue ergebnisse de primatologie. Fischer, Stuttgart.

Thorington, R. W., Jr. 1968. Observations of squirrel monkeys in a Colombian forest. Pages 69–85 *In* L. A. Rosenblum and R. W. Cooper, eds. The squirrel monkey. Academic Press, New York.

Thorington, R. W., Jr. 1970. Feeding behavior of nonhuman primates in the wild. Pages 15–27 *In* R. S. Harris, ed. Feeding and nutrition of nonhuman primates. Academic Press, New York.

NEOTROPICAL PRIMATES: ASPECTS OF HABITAT USAGE, POPULATION DENSITY, AND REGIONAL DISTRIBUTION IN LA MACARENA, COLOMBIA

Lewis L. Klein *and* Dorothy J. Klein

INTRODUCTION

While searching between July and November 1967 for a suitable area in which to make a year-long intensive study of spider monkey social organization, we collected limited amounts of information on the distribution of primate taxa in selected parts of La Macarena National Park, Colombia, and a few adjoining areas. These distributional data and our intensive observations of feeding behavior and habitat usage made at a single location form the basis for several hypotheses about regional primate distribution and density.

Much of the initial surveying and our subsequent study occurred within the boundaries of La Macarena National Park, which in 1968 was still a national reserve (Figure 1). The park is situated approximately 3° north latitude between 73° and 74° west longitude and comprised about 11,000 km² (4,250 sq miles). It consists of three types of terrain: an isolated mountain range, foothills, and the floodplains of the three major rivers and several smaller tributaries. The mountain range, with maximum elevations of about 3,000 m (12,800 ft) runs roughly parallel to the Cordillera Oriental. It has a predominantly north–south axis about 125 km (78 miles) long, and an east–west width of about 25 km (16 miles). The foothills on the eastern slope of the sierra are relatively wide and merge gradually into floodplains at elevations of 250–350 m (820–1,150 ft) above sea level to the south. In contrast, most of the western slope of the mountains is precipitous.

Most of the mountainous and foothill regions of the park is covered by forest presumed to be inhabited by nonhuman primates. However, we were unable to investigate these areas except for several very brief excursions into the foothills, as discussed below.

The remaining two-thirds of the park's area consists of floodplains, primarily of three rivers—the Guejar, Ariari, and Guayabero—which in 1968 formed the major natural boundaries of the forest reserve to the north, south, and east.

The study site and most of the areas surveyed were located on or near these floodplains, which were inundated either yearly or at least once every several years. The forests found at these inundated locations were not floristically uniform, despite their extensive continuity, taxonomic heterogeneity, and freedom from slash and burn agricultural practices. For example, in our study site of 780 ha (3 sq miles) on the north bank of the Guayabero, using gross criteria of presence and abundance of several easily identifiable trees, we were able to identify eight different types of forest communities. The existence of these small, spatially distinct floristic communities is supported by the work of some tropical foresters (Sawyer and Lindsey, 1971; Rosayro, 1958). In our region the identification of such communities was frequently used by natives in deciding where to plant crops.

The patchwork of forest communities probably results from varying durations of inundation, varying amounts of silt deposition or soil removal, changes in drainage patterns ranging from blockage of small seasonal streams to major changes in the riverbed, several types of seral succession, and chemical composition and porosity of the soil.

70

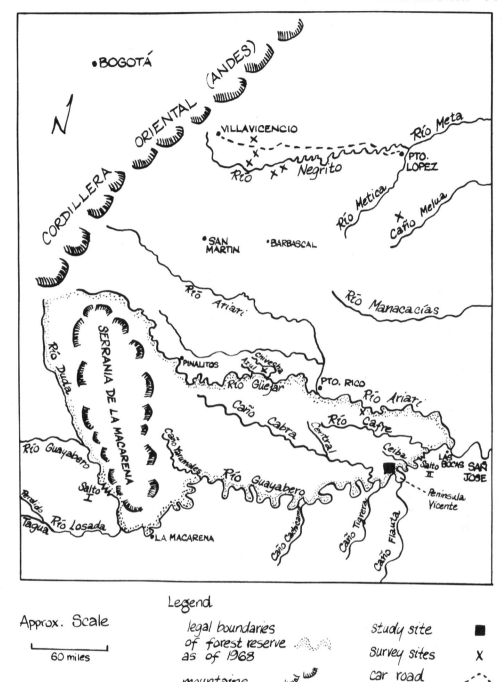

FIGURE 1 Map of La Macarena region showing location of study site and areas of preliminary survey.

The most interesting aspect of these separable forest communities was the way in which they appeared to differentially affect the seasonal and daily movements and feeding of several of the primate taxa. Our data on these matters is most adequate for *Ateles belzebuth* (approximately 650 hours of observation). Contrasting aspects of habitat usage will be outlined with comparative, but considerably less, material on sympatric populations of *Cebus apella, Saimiri sciureus, Alouatta seniculus,* and nearby populations of *Lagothrix lagotricha.* We feel that further delineation of some of these taxonomically correlated patterns of habitat usage may explain some of the larger-scale anomalies of taxonomic diversity and population densities noted in several of the areas explored outside the specific study site.

FOREST ASSOCIATIONS AND HABITAT USAGE

Eight forest types were identified at the study site: (1) heterogenous forest on creek and levee banks, (2) *Chrysophyllum* mixed forest, (3) "typical" study site mixed forest, (4) *Brosimum* swamp, (5) lakeside, shrub-tree association, (6) *Cecropia* stands, (7) clear tree swamp, and (8) aerial root swamp.

The first three of the eight types of forest were considerably more diverse than the others with respect to tree species present. About 40–60 percent of the study site area was covered by these three formations. In these associations crown continuity between adjoining trees was well developed, generally occurring at several levels from 15–31 m (50–100 ft) from the ground. Except for very scattered and occasional emergents growing to heights of 40–43 m (130–140 ft) (usually *Ficus* sp. or *Ceiba* sp.), the tallest trees were about 34 m (110 ft) high. Four types of tall palm trees were common but scattered—two were associated with one type of diverse associations and two were found in the other two diverse forest types. In general, the understory of these three associations was never thick enough to seriously impede walking and was composed predominantly of saplings. Diverse forests of these three types were inundated in 1968 for periods ranging between 1 week and 2 months. The three associations were distinguished from one another on the basis of absence or presence of specific varieties of palms and several large conspicuous trees, frequently used by *Ateles*, e.g., *Chrysophyllum*.

The most botanically heterogeneous forest communities were found on the highest points within the study site, which were probably the remnants of natural levees. In 1968 most of these habitat types were inundated for 2 weeks or less, and in many years they may not have been flooded at all. The relatively greater diversity of trees in this community, when compared to the other two types of taxonomically diverse forests, is probably the result of a combination of the slightly higher elevation and more rapid drainage.

The heterogenous forest on creek and levee banks comprised less than 5 percent of the study site area, yet included those areas in which contacts with *Ateles* occurred for a longer period of time throughout the year (from February through September) than in any other comparably sized forest sector. Levee forests, however, were used most intensively when certain trees, commonly found only there, were producing ripe fruit, particularly a species of *Hyeronomia*. About 50 percent of contact time with *Ateles* during the February ripening of *Hyeronomia* occurred in these communities. Although this botanical assemblage was important for spider monkeys throughout the year,

marked and radical fluctuations in usage occurred. For example, although we spent much time looking for spider monkeys along the higher creek beds in October and November 1968, when most of the rest of the forest was flooded and inaccessible, we logged less than 10 percent of total observation time with *Ateles* in this habitat. About one-third of our contacts with capuchin and squirrel monkeys in October and November occurred in forest communities on creek and levee banks.

The infrequent utilization of the heterogenous forests by *A. belzebuth* in October and November was correlated with an enormous increase in the proportion of daylight time they were observed feeding, resting, and moving among four discontinuous botanical communities composed of clustered concentrations of an unidentified species of *Brosimum*, locally known as "guaymero." *Brosimum* trees, which reached 46 m (150 ft) in height, were among the tallest in the region. They grew in association with several species of *Inga*, with less frequently occurring specimens of *Calophyllum* sp. and with an occasional specimen of an enormous spreading banyan-type fig, locally called "chivecha." At the edges of these communities, there were commonly clusters of a small palm (*Bactris* sp.), which is less than 12 m (40 ft) in height. This was the only palm directly associated with clusters of *Brosimum* at the study site. Collectively, the four major *Brosimum* communities comprised between 18–32 percent of the study site. These areas were inundated for about 4 months in 1968, but drained rapidly toward their peripheries as the river fell.

The marked shifts in area usage and arboreal travel routes coincided with a small but incomplete subsidence of water during the months of September and October and the ripening of *Brosimum* (late September and October) and *Calophyllum* (October). These fruits comprised about 70 percent of *Ateles* diet at that time (Klein, 1972). Approximately 60 percent of our observations of spider monkeys during this period occurred in *Brosimum* swamp communities. This compares with less than 10 percent of the observations during the preceding month and less than 5 percent of the observations from February, March, and April. Even when not bearing ripe fruit, the larger *Brosimum*, *Calophyllum*, and *Ficus* trees were occasionally used as resting and sleeping sites. Although a comparable quantitative estimate for the use of *Brosimum* swamp formations by sympatric primate taxa is impossible to provide, our contacts with other species in these areas during September and October were markedly less frequent than with *Ateles*.

Several other types of botanical formations occurred on the terrain, which was inundated for lengthy periods of time in 1967 and 1968. Some of these, in contrast to *Brosimum* swamp, were rarely if ever used by *Ateles*. Along some of the borders of the oxbow lake El Tigre and its connecting channel with the river, there were several pure stands of *Cecropia,* and adjoining areas were covered by relatively short, willow-shaped trees. These areas comprised only a small part (about 2–4 percent) of the study site and were inundated from 5 to 8 months during 1967 and 1968. Although *Ateles belzebuth* was neither seen nor followed to these relatively homogenous *Cecropia* stands, *Alouatta seniculus* and *Saimiri sciureus* were. Moreover, both *A. seniculus* and *Lagothrix lagotricha* were seen frequently in similar areas on the south bank of the Guayabero opposite the study site and *Ateles belzebuth* were not. The summed observation time for these three primate taxa was less than 20 percent of the approximate 650 hours of observations logged for *Ateles.*

"Clear tree swamp," another habitat type on the seasonally flooded plains, comprised an estimated 12–24 percent of the study site area. These areas were characterized by being under water for long periods of time (approximately 6 months) and by becoming deeper towards their centers, forming a catch basin for stagnant water. Maximum tree height appeared to be 24–31 m (80–100 ft), but the crowns of individual trees were often massive in diameter. Palm trees were absent except at the boundaries, where stands of small, spiny-trunked *Bactris* sometimes occurred. Although the penetration of sunlight to the forest floor was relatively high in these areas, understories of either saplings or herbaceous vegetation were usually absent or sparse. However, small patches of grasses occasionally occurred. To the observer, these forests appeared to be neither regenerating nor in the process of being replaced by another type. It is suspected that this open type of swamp forest may have developed on terrain where relatively recent changes in drainage patterns adversely affected trees that were already present. We noted little fruit on these trees in the months we were able to view them. What little was seen appeared to be the fruit of epiphytes or of an occasional isolated tree growing on small elevated hillocks. Consequently, while spider monkeys were frequently followed into and through these open swamps, and were frequently observed using trees on the border for resting, they were rarely observed to feed there. In contrast, we saw *Saimiri* and *Cebus* in these communities frequently foraging for insects, and on one occasion *Cebus* was observed eating what were probably frogs' eggs, which were suspended from low branches overhanging the standing water.

The last of the identified forest communities, aerial root swamp, was defined primarily on the basis of a single structural feature—the presence of aerial roots. Communities of this type, which were inundated for periods as long as 5 months, comprised about 3–9 percent of the study site area and were characterized by large numbers of short, 15 m (50 ft) or less, stands of trees with extensive aerial–pneumatophoric roots and lianas. As with *Cecropia* stands, spider monkeys were neither encountered nor followed to these areas within the boundaries of the study site, although *Saimiri* and *Cebus* were.

PRIMATE DISTRIBUTION DATA

Our preliminary explorations took us to several locations within or just outside the park's boundaries. These included: (1) an excursion by canoe and foot between September 24 and October 4, 1967, from the mouth of Caño Cabra upstream to its junction with Caño Central—2 days were spent in a foothill area above the floodplain; (2) 4 days within forests bordering the Losada River; (3) 7 days between Barranco Colorado on the Ariari River and the headwater areas of the Cafre River; (4) 5 days on the north bank of the Guejar River between Caño Azul and Caño Chivecha; and (5) a total of 9 days at several sites on the south and north banks of the Guejar River downstream from the settlement of Pinalitos. Throughout the study period, we also made occasional short trips into forests on the south bank of the Guayabero River between and bordering its tributaries La Flauta and La Tigrera.

Many of the arboreal communities occurring at the study site also appeared to occur at most of the sites explored briefly. However, in some areas certain communities, e.g., *Brosimum* swamp, appeared to be either absent, curtailed in size, or infrequent; and at several locations trees and vegetational structuring not seen at the study site were noted. In some instances these floristic differences appeared to be associated with either the presence of primate taxa in addition to the four diurnal species at the study site or, in one case, a possible replacement. The apparent replacement was noted in two areas—the upstream noninundated sections of forests bordering Caño Cabra (2 days of observation) and an area near the headwaters of Caño Cafre (1 day of observation). In both these areas *Callicebus moloch,* an animal not present at the study site, were observed. These observations along with several other indications in the literature (Mason, 1971), suggest a degree of competitive displacement between *Callicebus* and *Saimiri.* The fact that those forests in which we observed *Callicebus* and not *Saimiri* were composed of trees no taller than 22 m (70 ft) may have had some bearing on finding *Callicebus*

74 KLEIN and KLEIN

but not *Saimiri*. Extensive and continuous tracts of poorly developed forests did not occur at the study site. Similarly, *Callicebus torquatus* were observed in the vicinity of the rapids, Angostura I (Salto), and reported to be present elsewhere on the south bank of the Guayabero where they occurred on hillside terrain. There, the trees were shorter than the general average for the region and were similar to those in which *C. moloch* were seen on the north bank foothills. An interesting complication is added by the reported presence of *C. moloch* in certain areas of the south bank. However, they were limited to extensively inundated sections of forest, which sometimes supported well-defined stands of trees substantially lower in height than those on adjoining, better-drained terrain.

The presence of one species of *Callicebus* on the south bank of the Guayabero and the probable presence of a second was indicative that the south bank forests everywhere supported a greater variety of primate taxa than did the north bank areas. *Lagothrix lagotricha, C. torquatus,* and probably *C. moloch* occurred on the south bank, in addition to the four diurnal primate taxa occurring at the study site. Woolly monkeys were seen on the south bank from Angostura II as far upstream as the forested environs of La Macarena village and frequently and regularly in the forest across the river from our campsite. Yet they were neither seen nor heard on the north bank between the two Angosturas. We were also told by native observers that they occasionally encountered additional primates in upstream areas of the south bank tributaries La Tigrera and La Flauta. Their descriptions best fit *Cacajao melanocephalus* and *Cebus albifrons*. Unfortunately, we were unable to confirm these reports. Neither of these species was ever observed at the study site. Thus, on the south bank there may have been as many as 10 primate species; five in addition to the four diurnal and one nocturnal taxa resident at the north bank study site.

A simple explanation for the greater diversity on the south bank would be that the river acts as a barrier to dispersal. This, however, can only be a partial explanation. First, the width of the river bed is only 137 m (150 yd) at the study site. Second, floating vegetation in masses that would easily support the weight of monkeys was frequently seen during certain seasons moving both down and across the river. Thirdly, between La Macarena village and Angostura II, the Guayabero River meanders extensively, resulting in occasional oxbow lake formation and shifts of large sectors of bankside forest, 259–518 ha (1–2 sq miles) from one side of the river to the other (see Figure 2). Areas this large could probably support small groups of primates for periods sufficient for the changeover to become completed without necessitating actual rafting or swimming by individual animals.

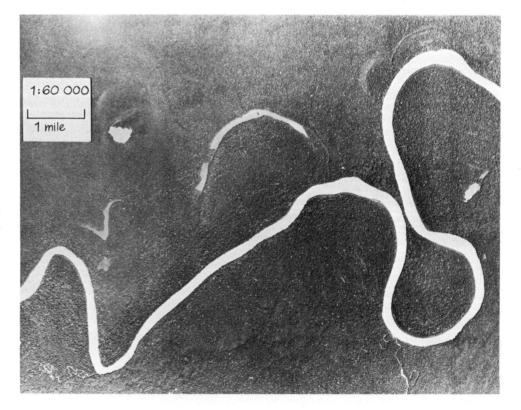

FIGURE 2 Aerial photo of study site forest, Laguna El Tigre, and south bank forest between Caños La Flauta and La Tigrera. Source: Instituto Geografico "Agustin Codazzi," Bogotá, Colombia, taken in 1961.

The greater variety of primate taxa on the south bank of the river appears to be related to the greater heterogeneity of forests on the south bank. For example, along and between Caño Flauta and Caño Tigrera, there were hillsides and large sectors of undulating and level terrains that were high enough to escape flooding in most, if not all, years. These areas supported well-developed, tall, continuous, and taxonomically diverse forests. These forests existed in addition to the major types of arboreal forests identified at the study site, some of which were described above. An analysis of the distribution of primate genera in the area of La Macarena should probably minimize the importance of rivers as dispersal barriers and maximize the importance of botanical diversification and ecologically competitive interactions between differing taxa.

POPULATION DENSITY AND GROUP SIZES

Adequate estimates of the study site population density of *Ateles belzebuth* were obtained. The validity of the density estimates for primates other than *A. belzebuth*, however, include a large magnitude of probable error. They are not based upon fully defined annual home ranges, nor positive reidentification of the same troops or individuals encountered on separate occasions. However, they represent estimates of population density in a virtually undisturbed forested floodplain habitat and, as such, may be of some value to future students or to researchers working with populations in other locations. The group size estimates for all taxa (Table 1) should be fairly reliable. With one exception, specified below, they are based entirely on minimal contact periods of 15 minutes duration with the group counted.

Saimiri sciureus. The population density of squirrel monkeys was estimated to be 50–80 animals per square mile (259 ha). The most satisfactory counts of fairly cohesive bisexual groups resulted in figures of 25–35 independently locomoting animals. Between April and June group size usually increased by the addition of 5–10 infants. On two occasions temporarily isolated male squirrel monkeys were observed following subgroups of spider monkeys. Between three and six separate groups of *S. sciureus* ranged over the approximately 780-ha (3 sq miles) study site.

Cebus apella. The population density of *C. apella* was estimated to be 15–25 animals per square mile (259 ha). Counts of animals ranged from a single instance of an individual to groups composed of 12 independently locomoting capuchins. Median group size was approximately six animals. The number of groups at the study site was estimated to be between four and six.

Alouatta seniculus. The estimated population density for *A. seniculus* at the study site is low relative to the existing data for howlers from other areas (Chivers, 1969; Neville, 1972). There appeared to be between 30 and 75 animals per square mile (259 ha). Aside from several observations of either isolated males or pairs of isolated males, groups ranged in size between three and six independently locomoting individuals, with a median of 4.3. If infants are included, i.e., animals carried by females at most tree crossings, the range of group sizes was from three to eight, with a median of almost five. At least 75 percent of the bisexual groups contained only one fully adult male. There were at least 6 groups of howlers at the study site, but probably no more than 15. Shifts in area usage appeared to be at least as marked as those of the spider monkeys, but were not concordant with *A. belzebuth* throughout the year. Including isolated males, it was estimated that population structure consisted of 28 percent adult males, 40 percent adult females, 22 percent juveniles, and 10 percent infants. The comparable figures for *A. palliata* on Barro Colorado Island in 1967 (Chivers, 1969) were 22 percent adult males, 41 percent adult females, 20 percent juveniles, and 17 percent infants; for groups of *A. seniculus* on a ranch in Guarico, Venezuela, 29 percent adult and subadult males, 33 percent adult and subadult females, 21

TABLE 1 Estimated Study Site Population Densities, Group Size and Number of Groups Present

	Estimated Population Density per Square Mile	Range of Group Size[a]	Estimated Number of Groups Utilizing 780 ha (3 sq miles) Study Site
S. sciureus	50–80	25–35	3–6
C. apella	15–25	6–12	4–6
A. seniculus	30–75	3–6	6–15
A. belzebuth[b]	30–40	17–22	3

[a] Independently locomoting animals.

[b] See text for methods of determining total group size from subgroup data.

percent juveniles, and 18 percent infants (Neville, 1972).

Ateles belzebuth. Although our estimates of population density, group size, and composition are much firmer for this species than those just given for *Alouatta, Saimiri,* and *Cebus,* they were, of necessity, based on different techniques of censusing. Simply counting observed groups of spider monkeys was of little value in determining either population density or total group size. Groupings of spider monkeys, in comparison to other sympatric primate taxa, were markedly labile and flexible. Counts ranged from isolates of both sexes and mothers with infants (15 percent of 498 counts: 3 percent adult male, 10 percent adult female, and 2 percent adult female with infant) to groups composed of 22 independently locomoting individuals (see Table 2). Counts of four or more independently locomoting animals also revealed no adult males, one adult male, or two or more adult males (see Table 3). Although median number of independently locomoting spider monkeys per count over the year was 3.5, it varied monthly from a high of 5.6 in May to a low of 1.5 in December (see Table 4).

This quantitatively complex and variable picture can be used to determine population density, group size, and group composition only if the reliable identification and reidentification of most individuals is feasible. Fortunately, *A. belzebuth* has a considerable amount of intrapopulational variability in facial and clitoral pigmentation (see Figure 3). In time, despite the lability and flexibility of grouping tendencies and relative absence of social tendencies resulting in fixed limits to interindividual spatial dispersals, almost all *A. belzebuth* at the study site could be assigned to one of three different and mutually exclusive social groups whose members were rarely assembled in a single location (Klein, 1972). The larger of the two best-

TABLE 2   The Percentage of Subgroups Consisting of One to Eight or More Independently Locomoting *A. belzebuth*

Subgroup Size	Frequencies	Total Subgroups (%)
1	75	15
2	105	21
3	72	14
4	78	16
5	36	7
6	36	7
7	20	4
8–22	76	15
TOTAL	498	99

TABLE 3   Composition of Subgroups of Four or More Independently Locomoting *A. belzebuth*

	Frequency (%)	Range of Subgroup Sizes
Bisexual subgroups		4–22
Bisexual with two or more adult males	76 (33)	4–22
Bisexual with one adult male	76 (33)	4–10
SUBTOTAL	152 (66)	
Single sexed groups		4–11
Entirely male	2 (1)	4
Entirely female with no dependent young	13 (6)	4–5
Entirely female with dependent infants or juveniles	13 (6)	4–8
Entirely female with and without dependent young	49 (21)	4–11
SUBTOTAL	77 (34)	
TOTAL:	229 (100)	

known groups contained 22 independently locomoting animals and 5 infants in February 1968; the smaller, 17 independently locomoting animals and 3 infants in September 1968. Five fully adult males were members of the larger group; 3 fully adult males were members of the smaller. The ratios of adult males to adult females were 1:2.4 and 1:3.5, respectively. Annual home range was estimated to be on the order of 1–1.5 sq miles (259–388 ha), with overlap roughly 20–30 percent. The study site population density of *A. belzebuth,* including the third group, whose members were only rarely encountered, was consequently estimated to be 30–40 independently locomoting animals per square mile (259 ha).

TOTAL PRIMATE DENSITY AND COMPARISONS

Total primate density at the study site, excluding *Aotus,* was probably 125–250 individuals per square mile (259 ha). Although no comparable estimates can be made for the other areas visited, there were some indications that spider monkey density on the south bank of the Guayabero was considerably lower than on the north bank study site. They were entirely overlooked on the south bank by resident Indians. We saw *A. belzebuth* on the south bank only once, and heard them on only a few other occasions in areas in which *Lagothrix lagotricha* and *Alouatta seniculus* were quite frequently seen and heard. This strengthens our

TABLE 4   Intermonth Variations in Subgroup Size of All *A. belzebuth* Observed at Study Site

Months	Median	Mode	Range	Number of Encounters	Encounters with Isolated Individuals (%)	Encounters with Subgroups of Eight or Larger (%)
Jan.	2.9	3	1–17	64	17	2
Feb.	4.0	2	1–18	103	5	16
Mar.	4.3	4	1–10	48	4	12
Apr.	3.7	2	1–11	82	12	15
May	5.6	3	1–20	43	7	40
June–Aug.	2.7	1	1–6	19	32	0
Sept.	2.2	2	1–11	47	28	6
Oct.	5.2	1	1–22	44	21	43
Nov.	2.2	2	1–11	36	25	3
Dec.	1.5	1	1–4	12	50	0
TOTAL	3.5	2	1–22	498	15	15

earlier suggestion that botanical diversity, ecological competition, and forest productivity must be incorporated into explanations accounting for geographical distribution, taxonomic diversity, and population densities of single taxa.

NEED FOR CONSERVATION

The effects of human activities on nonhuman primate populations in La Macarena may be of interest. We made no specific investigations of this subject, but it

FIGURE 3   *A. belzebuth* pausing subgroup. Note variations in facial pigmentation.

was a matter of concern when looking for a study site. Several areas were rejected because of forest felling and/or evidence of heavy hunting pressures. The areas most affected were those abutting on the boundary rivers, particularly the higher banks less likely to be inundated. Virtually all suitable locations along the Ariari and Guejar rivers had small homesteads or clearings. In comparison, settlement along the Guayabero was relatively light, presumably because of the difficulties involved with transporting cash crops past the two rapids; nevertheless, extensive areas had been felled below Angostura I and within 50 km of the village of La Macarena (Grimwood, 1968). However, the effects of human habitation on wild primates in the La Macarena area were not uniform. Many areas had been colonized only briefly, and the economic and agricultural practices appeared to have differential effects upon the resident taxa.

Hunting, for example, appeared to seriously affect *Ateles* and *Lagothrix*. They are a favorite food, since they are large and relatively easy to shoot. For many settlers, particularly in their first few years of colonization, they provided a major source of protein. Unfortunately, from the conservation point of view, *Ateles* is a slow-breeding animal and *Lagothrix* may be as well. Spider monkey females, for example, do not appear to give birth before their fifth year, and thereafter only once every 2 years. One recent colonist reported killing 22 spider monkeys in a 2-month period. He claimed it was necessary or he would have starved, since he had lost most of his crops during the 1967 flooding. On the other hand, the elimination of animals such as jaguars, ocelots, and some of the larger snakes, as a consequence of the trade value of their skins at that time, probably had a minimal positive effect on the larger primates such as *Lagothrix* and *Ateles* and may have had a greater positive effect on primates such as *Saimiri* and *Cebus,* which were not usually shot for food.

Systematic trapping for primates, as far as we could determine, was not practiced in the park. Captives seen in rancherias and the markets of Villavicencio, which originated from La Macarena or nearby areas, were usually obtained as a result of shooting females with young.

Agricultural activities in the area may also have had both positive and negative effects. Forest was felled selectively and generally in small plots; this was done in part to facilitate claiming maximum amounts of land and in part to allow the planting of a diversity of crops. House sites and sites on which fruit trees and yucca were to be planted were usually located on the highest banks and areas. As noted above, many of these terrains supported the most heterogenous portions of

the forest—areas intensively used by most of the primates, particularly *Ateles.* For crops such as bananas and corn, sites with a slight inundation were selected. This again resulted in the destruction of forest types that, at the study site, were highly favored by spider monkeys. Less-productive forest areas, e.g. clear tree swamps, were not favored areas for cultivation. Our general impression was that the degree and nature of clearing that occurred in most park areas probably had lesser effects upon taxa such as *Saimiri* and *Cebus,* who were more likely to be distributed over different types of forest than were *Ateles.* Differences in locomotion and feeding habits, in combination with the practice of felling discontinuous patches of forest, were also likely to interfere more seriously with the travel routes of spider monkeys than with those of squirrel and capuchin monkeys.

In the La Macarena region, *C. apella* was the only primate able to turn agricultural activities to immediately positive advantage by raiding crops, particularly maize.

ACKNOWLEDGMENTS

The field research reported in this paper was financed primarily through predoctoral fellowship 1 F1 MH-31, 722-01, and a supplementary research grant, 1 RO4 MH 13608-01, provided by the Department of Health, Education, and Welfare, Public Health Service, National Institutes of Health.

REFERENCES

Chivers, D. J. 1969. On the daily behaviour and spacing of howling monkey groups. Folia Primatol. 10:48–102.

Grimwood, I. R. 1968. Reports and recommendations on the Sierra Nevada de Santa Marta National Park, the Isla Salamanca National Park, the Tairona National Park, the La Macarena National Reserve. British Ministry of Overseas Development.

Klein, L. L. 1972. The ecology and social organization of the spider monkey, *Ateles belzebuth.* Unpublished Ph.D. Thesis. Univ. of California, Berkeley.

Mason, W. A. 1971. Field and laboratory studies of social organization in *Saimiri* and *Callicebus.* Pages 107–137 in L. A. Rosenblum, ed. Primate behavior 2. Academic Press, New York.

Neville, M. K. 1972. The population structure of red howler monkeys (*Alouatta seniculus*) in Trinidad and Venezuela. Folia Primatol. 17:56–86.

Rosayro, R. A. de. 1958. Recent advances in the application of aerial photography to the study of tropical vegetation. *In* Proceedings of the Symposium on Humid Tropics Vegetation. Publication of the Unesco Science Cooperation Office for South East Asia.

Sawyer, J. O., and A. A. Lindsey. 1971. Vegetation of the life zones in Costa Rica. Indiana Academy of Science, Monograph No. 2. 214 pp.

NOTES ON THE ECOLOGY AND BEHAVIOR OF THE PYGMY MARMOSET (*CEBUELLA PYGMAEA*) IN AMAZONIAN COLOMBIA

Martin Moynihan

INTRODUCTION

Tamarins and marmosets are an abundant and successful group of New World primates. They are "monkeys," members of the suborder Anthropoidea, and distinguished by small size and a series of correlated characters. Within this group, the smallest, the pygmy marmosets of the genus (or perhaps subgenus) *Cebuella,* are remarkable for their extreme specialization. They are, in a sense, the culmination of a major evolutionary trend.

It is hoped that the following brief account of the known or surmised behavior and ecology of free-living pygmy marmosets in the "wild" may illustrate and explain some of the peculiarities and constraints within which they have to operate.

DESCRIPTION

The coloration of *Cebuella* is largely tan grizzled or brindled with black. Although Hershkovitz (1968) considered the pattern to be primitive, it is also immediately functional (see below).

There is considerable variation in the tone of the underparts and the boldness of the brindled stripes on the back. Several coloration variants may be adapted to avoid detection against different backgrounds. The existence of such polymorphism may explain why farmers near El Pepino and Rumiyaco believe there are two species of pygmy marmosets. These people are acute observers, but I believe they are mistaken in this particular instance. They say that both types occur

in the same places and have the same habits. I have seen individual marmosets of distinctly different appearance in the same groups.

HABITAT

Cebuella seems to be confined to the upper part of the Amazon valley (Hill, 1957), a lowland tropical environment of generally humid and stable climate (dry seasons are seldom excessive or prolonged).

I was able to observe several groups of pygmy marmosets in and around the localities called El Pepino and Rumiyaco, between the towns of Mocoa and Puerto Asís in the Putumayo region of Colombia, during four field trips (September 17–24, 1968; July 4–11, 1969; February 19–23, 1970; August 12–14, 1970).

Under natural conditions, the Putumayo region probably would be covered by high and relatively open "monsoon" or "rain" forests. These forests would be interspersed with small patches and strips of low, dense vegetation, thickets, and tangles on poorer soils, in swamps, on rocky outcrops, at treefalls, and in edge habitats along streams and rivers.

Under present conditions, the human population of the region is increasing; much of the forest has been cut down; and there is the beginning of a pollution problem, a by-product of the developing petroleum industry.

It is not possible to determine the original habitat preferences of pygmy marmosets. They may well have occurred along edges of forests. According to the local

Indians (and also H. LeNestour, personal communication), they can still be found along edges of forests in the more remote and less disturbed parts of the Putumayo. They may also have occurred within climax forests or at least older communities, in association with trees of *Parkia* sp. (Hernández-Camacho and Cooper, 1976). However, I did not find them in either situation.

At present, all or most of the local marmosets of El Pepino and Rumiyaco have become commensals of man, adapted to some of the results of human activities, like some American monkeys of other regions (e.g., the tamarin *Saguinus geoffroyi* in Panama—see Moynihan, 1970). They seem most abundant in "hedges," strips and clumps of degraded woods found between pastures and crop fields from which the most economically valuable (tallest) trees have been removed by selective cutting and from which many of the larger mammals have been driven by hunting.

The local marmosets occur in small groups. Three to six individuals comprised each of the six groups that I could follow closely. In thick vegetation the animals are very difficult to count precisely, since they tend to "string out"; but settlers claim to have seen groups of approximately eight or nine individuals. This suggests that the basic social unit is the family, composed of an adult male and female with their young of the year, sometimes with older young of previous years with their own mates and offspring. If so, the social organization of the species is much the same as those of many other marmosets and the ecologically similar or related tamarins. The average size of groups may be smaller in *Cebuella* than in such tamarins as *Saguinus geoffroyi* and *S. fuscicollis*, but the difference is not great and may be due to recent human intervention.

Each group of pygmy marmosets has a well-defined and apparently persistent home range, often the whole of an isolated or semi-isolated hedge or patch of scrub, at times several hundred meters long. These home ranges may be territories defended against intruders, but the usual arrangement of hedges and patches is so dispersed that face-to-face encounters between groups must be infrequent.

FEEDING HOLES

The El Pepino and Rumiyaco marmosets eat a variety of fruits, buds, and insects. Local people catch them by baiting traps with any convenient fruit, even those not native to the area, such as bananas and plantains. The animals also come down to the ground or cleared pasture to catch grasshoppers. To reach the latter, they have been seen running across asphalt highways. This catholic selection of foods is typical of many monkeys. A more distinctive habit of the pygmy marmosets is "sap-sucking."

Every family or group of pygmy marmosets has one or more trees in its home range that are riddled with small holes, sometimes hundreds of them (see Figure 1). The majority of the ones that I inspected were roughly circular, approximately 1–1.5 cm (0.4–0.6 in.) in diameter and half as deep. Some holes appeared to be quite new. In these, it was evident that the cut extended through the bark and just down to the next level, presumably the cambium, but no further. Other holes appeared to be old. They were partly filled in by secondary proliferation of new bark, a sort of scar tissue spreading inward from the sides of the cut.

The marmosets visit the holes, at least the newer ones, very frequently and repeatedly. They spend hours going from hole to hole, usually staying no more than a few minutes or even seconds at each. When an individual reaches a hole, it puts its face or muzzle down into the cavity and follows this by slight but rapid movements of the head. A human observer watching from the back or side cannot analyze these movements in detail, but they seem to accompany vigorous action of the mouth and jaws, chewing, licking, or sucking.

Obviously, the marmosets are getting something

FIGURE 1 Pygmy marmosets on a feeding tree.

from the holes. They may obtain an occasional insect or grub, but there was no accumulation of insects in the holes I inspected. More significantly, the same holes may be visited at very short intervals, and the later visits are not briefer or simpler than the earlier ones. Insect feeding is not a likely explanation of this behavior. A more plausible alternative is that they are feeding on the liquids, presumably sap, which can be seen to leak from the cut surfaces. This hypothesis is supported by the findings of sap or plant gum in the stomachs of pygmy marmosets in another area of Amazonian Colombia (Hernández-Camacho and Cooper, 1976). Considering the time spent in visiting holes, sap would appear to be a major food source for the marmosets.

The local people of El Pepino and Rumiyaco believe that the marmosets cut the feeding holes themselves. Hernández-Camacho has observed some gnawing by marmosets. The fact that existing holes are used frequently and repeatedly indicates that the actual digging or cutting cannot be a very frequent procedure. I never saw a marmoset dig or cut a hole in a tree that was already riddled, but I did see one individual begin a hole in a previously untouched tree. It is conceivable that the marmosets may prefer to gnaw at sites where the bark has already been damaged by other animals, but their finished feeding holes cannot be confused with either the deeper excavations of woodpeckers (e.g., *Dryocoups* and *Campephilus* spp.) or the more extensive "stripping" operations of the local pygmy squirrel.

I found marmoset feeding holes in several different species of trees, including an *Inga* (probably *I. spectabilis*), *Matisia cordata*, and a tree named *cedro* by the settlers (possibly *Cedrela odorata*, Pérez-Arbeláez, 1956). I did not find (or recognize) holes in trees of *Parkia* sp., which appear to be favored by *Cebuella* elsewhere (Hernández-Camacho and Cooper, 1976) and which may be the original or "natural" source of sap for many animals in less-disturbed conditions. This seems to be another indication that the El Pepino and Rumiyaco marmosets have already made the translation or transferral to an extraneously managed environment.

All the holes that I saw were in trees of appreciable size, and presumably age, in trunks and large branches, from only a few centimeters to more than 12 m (40 ft) above ground. The lowest holes seemed older, while the higher holes looked progressively younger. The marmosets must begin drilling at the bottom and gradually work upward. I saw one tree that appeared to have been recently attacked. There were only a few holes in the trunk, all below 3 m (10 ft), in spite of the fact that the tree was moderately thick and tall, reaching a total height of approximately 10 m (30 ft). The marmosets also tend to go from bottom to top during single feeding visits.

To my knowledge, *Cebuella* is the first primate to be found to perform such elaborate sap-sucking. Several other species, prosimians as well as monkeys (see, for instance, Charles-Dominique, 1972), are known to eat plant gums. Some have evolved special adaptations of morphology and/or behavior to facilitate obtaining such food, but they apparently do not usually drill for it. A few *Callithrix geoffroyi* that I observed in the National Zoological Park in Washington, D.C., chewed on dead branches with quite unusual vigor and persistence. All species of *Callithrix* share with *Cebuella* the peculiar character of elongated lower incisor teeth, which may be an adaptation to gnawing hard materials.

MOVEMENTS

Pygmy marmosets are thoroughly diurnal. Like many diurnal and endothermic vertebrates in the tropics, they are more active in the cooler hours of the mornings and late afternoons than at midday. It seems that their periods of activity are more prolonged, on the average, than are those of their larger relatives. This is what would be expected of such tiny animals, with their high metabolic and food requirements.

Like all the smaller New World primates except *Saimiri*, pygmy marmosets sleep in tree holes. Their sleeping holes are not in the same trees as their feeding holes.

When they are active but not alarmed, their movements and locomotory patterns are conventional, reminiscent of such tamarins as *Saguinus fuscicollis*. They can run along horizontal or diagonal branches as if on the ground, with a galloping gait, yet may walk or pace when advancing slowly on a branch. They can make long and nearly horizontal leaps of a meter or more, but in many circumstances they are also "vertical clingers and leapers" (Napier and Walker, 1967). Even when not visiting feeding holes, they spend much of their time moving up and down tree trunks. They prefer to rest sitting up or clinging to a trunk in a vertical position. (See Figures 1 and 2.)

Although pygmy marmosets can become tame in special situations, they seem to be terrified of potential predators, more so than any of their relatives. One group living in a hedge at the side of a highway paid little attention to either human beings or passing traffic, not even the heaviest, noisiest, most brightly painted or illuminated trucks. However, other less-sophisticated individuals in less extremely aberrant circumstances were exceedingly shy and timid. They use many antipredator devices, all of which seem to be

FIGURE 2 Hostile patterns of pygmy marmosets.

designed to avoid attracting the attention of a predator rather than distracting him. Pygmy marmosets do not show the spectacular "mobbing" behavior of many tamarins. Instead, like squirrels they frequently dodge behind trunks and branches. They also have developed several other protective types of locomotion. Sometimes they move exceedingly slowly, making the movements difficult to detect, as in the case of sloths (Figure 2). More often, they advance in spurts, lizard-like alternations of dashes and frozen immobility. The immobility, however, can only be partial without losing its effectiveness. An alarmed individual may continue to turn its head in all directions, on the lookout for danger, without becoming conspicuous, since the head is so small, often smaller than surrounding leaves swaying in the wind. The coloration of the species is highly cryptic. When individuals pass from one tree to another, they almost always prefer to take a low route rather than a high one, thus keeping as far away as possible from the canopy and minimizing exposure to flying birds of prey. (Birds of prey tend to become rare in the immediate vicinity of human settlements, doubtless another advantage of such areas from the point of view of the marmosets.)

SOCIAL BEHAVIOR

The intraspecific social reactions of these animals are partly mediated by "displays," behavior patterns that are specialized to convey information. The repertory of *Cebuella* displays, as shown by captive adult and juvenile individuals studied in the laboratory of the Smithsonian Tropical Research Institute, on Barro Colorado Island, is a modified version of the basic "language" of all New World primates (Moynihan, 1967). It probably is most similar to the corresponding systems of the marmosets of the genus *Callithrix* (see Epple, 1968, and comments in Moynihan, 1970), but it is far from identical. Some diagnostic features, such as the apparent use of purely "ultrasonic" alarm calls, may be further adaptations to avoid attracting the attention of predators (the higher the frequency, the less penetration of the sound). In the field, pygmy marmosets are very quiet within the range of frequencies audible to human ears, by far the quietest of the monkeys with which I am familiar.

RELATIONS WITH OTHER SPECIES

Pygmy marmosets must compete with many other animals. Competitors for food, and perhaps perching sites, include lizards and many birds, e.g., iguanas, tanagers, saltators, flycatchers, woodpeckers, and woodcreepers. The marmosets do not react to these animals in obvious manners, except for flinching when a bird comes too close.

In the artificial hedges, pygmy marmosets are largely segregated from other nonhuman primates. Although there were reports of groups of *Saguinus fuscicollis* traversing *Cebuella* ranges, I did not see any contacts between the two species, nor are they likely to be common. *S. fuscicollis* usually prefers treefalls and tangles *within* older forests. I might add that I watched *fuscicollis* and two other tamarins, *Callimico goeldii* and another form of *Saguinus,* possibly *graellsi,* at some length in other areas without seeing a trace of sap-sucking. In the upper Amazon, among the local primates, sap-sucking seems to be a "trick" of marmosets alone.

The pygmy marmoset seems to have come to a special arrangement with the local pygmy squirrels, probably *Microsciurus (flaviventer?) napi* (name from Hernández-Camacho, personal communication). These squirrels are widespread in the Putumayo. Like some of the tamarins, they prefer deep or old forest, but they are also regular and persistent occupants of second-growth hedges. In hedge-type habitats, they appear to be less common than the marmosets and less clustered (I saw only single individuals and pairs). The home ranges or territories of the squirrels and marmosets broadly overlap in some places. However, I never saw any direct encounters between individuals or groups of the two species. Both are abundant enough to suggest that this absence of personal contact is not coincidental. Probably some short-term avoidance mechanism is involved. The two species can repeatedly occur at exactly the same sites on the same day, but apparently never at exactly the same time.

The squirrels may have a varied diet. I watched only one aspect of their feeding behavior—bark-stripping. They often gnaw or chew off rather long, irregularly shaped, strips of bark and then appear to eat the bark itself. This bark-stripping is pertinent to the marmosets and can lead to remarkable spatial patterning. One graphic pattern was revealed by inspection of a large tree, *Inga* sp., near Rumiyaco.

The tree split into four equal trunks a meter or so above its base. Three of the trunks were covered with the marks of gnawing by pygmy squirrels. (I saw the animals at work.) Two of these trunks had no pygmy marmoset feeding holes in them; the third had a few holes on the lower part. The fourth trunk was dotted with marmoset holes all over and had only a very few squirrel marks. Obviously, the two species were dividing the tree between them. The number of marmoset holes was about as great as usual in a feeding tree of the species. This would suggest that the marmosets had taken what they wanted and left the rest. If the short-term avoidance mechanism is unilateral, this could mean that the squirrels are more likely to keep away from the marmosets than vice versa.

The competition between the two species is serious. Neither the drilling of the marmosets nor the stripping of the squirrels is good for the trees on which they are practiced. The actions of each species tends to have detrimental effects upon the other, at least insofar as they damage or impair the chances of survival of an essential or useful food source.

The social segregation of the two species, which permits or facilitates their competitive coexistence in the same places, seems to be partly a matter of timing. The local Indians state, and my own observations tend to confirm, that the squirrels remain active in the middle of the day, when the marmosets are resting, and also during rainstorms, when the marmosets often take shelter.

PROSPECTS

Cebuella faces a dubious future. The recent partial clearing of land by immigrant human settlers appears to have favored marmosets by providing new foods and patches of suitable habitats. Perhaps they are more widely distributed now than in the past. There are other human activities, however, that could prove to be disastrous.

The marmosets are subject to intense hunting and collecting pressures. The local people of the Putumayo find them entertaining. There are not many other equally attractive pets that a schoolboy can keep on his wrist or in his shirt pocket. The marmosets probably could withstand pressures of local origin. However, they cannot support in the long run an international traffic. They are being trapped for sale and export in large numbers with enormous mortality.

Official export figures are probably unreliable for animals as small and easy to smuggle casually as pygmy marmosets. Most trapped individuals may die before they can be shipped, but it is my impression that *Cebuella* is the monkey most commonly hawked about the streets of many cities of the Amazonian region. The trade should not be allowed to continue. It can only be regulated from the receiving, not the forwarding, end. The simplest solution would be to build up breeding stocks in zoos, laboratories, compounds, etc., to satisfy whatever demands may exist or develop. If this could be done, the species should survive, even expand and flourish, in the field, as well as in captivity, for a very long time.

SUMMARY

Cebuella pygmaea is the smallest and one of the most specialized of New World primates. Individuals of the species were observed near El Pepino and Rumiyaco in the Putumayo region of Colombia. There they have become commensals of man. They occur in family groups in hedges between pastures and crop fields. They eat a variety of animal and vegetable foods, but tend to concentrate upon sap-sucking. Many of their other distinctive features—locomotory habits, color, ultrasonic vocalizations—would appear to be anti-predator adaptations. Birds of prey probably were their most important predators under natural conditions. They have many competitors. They are usually segregated from *Microsciurus* sp. by a short-term avoidance mechanism. The partial clearing of forest by human settlers has been favorable to them in some ways. Under the present circumstances, however, they are in danger from excessive hunting for the export trade. If this can be controlled, they should continue to flourish indefinitely.

ACKNOWLEDGMENTS

My debt to many inhabitants of the Putumayo region must be obvious. I am particularly grateful to Sr. Jorge Fuerbringer and his family for all kinds of information, advice, and hospitality. Dr. Jorge Hernández-Camacho was kind enough to share with me some of the data that he has accumulated through his extensive field experience and to check and correct my more tentative identifications.

REFERENCES

Charles-Dominique, P. 1972. Ecologie et vie sociale de *Galago demidovii* (Fischer 1808; Prosimii). Advances in Ethology; J. Comp. Ethol. Suppl. 9:7–41.

Epple, G. 1968. Comparative studies on vocalizations in marmoset monkeys. Folia Primatol. 8:1–40.

Hernández-Camacho, J., and R. Cooper. 1976. The nonhuman primates of Colombia. This volume.

Hershkovitz, P. 1968. Metachromism or the principle of evolutionary change in mammalian tegumentary colors. Evolution 22:556–575.

Hill, W. C. O. 1957. Primates. III. Pithecoidea, Platyrrhini. Edinburgh Univ. Press.

Moynihan, M. 1967. Comparative aspects of communication in New World primates. *In* D. Morris, ed. Primate ethology. Weidenfeld and Nicolson, London.

Moynihan, M. 1970. Some behavior patterns of Platyrrhine monkeys. II. *Saguinus geoffroyi* and some other tamarins. Smithson. Contr. Zool. 28:1–77.

Napier, J. R., and A. C. Walker. 1967. Vertical clinging and leaping—a newly recognized category of locomotor behaviour of primates. Folia Primatol. 6:204–219.

Pérez-Arbeláez, E. 1956. Plantas útiles de Colombia. Camacho Roldán, Bogotá.

THE NONHUMAN PRIMATE TRADE IN COLOMBIA

Ken M. Green

INTRODUCTION

During the past several years, parameters of the primate trade have been discussed in detail (Cooper, 1968; Harrisson, 1971; Roth, 1969a,b; Thorington, 1972; Middleton *et al.*, 1972). It is the purpose of this paper to supplement this information with descriptions of (1) the present internal primate supply chain as it exists in the northern coastal region of Colombia around Barranquilla, (2) the existing Colombian regulations and legislation concerning primates, (3) past and present primate population pressures, and (4) the feasibility of primate conservation and management within Colombia. The author collected this information while serving with a Peace Corps primate conservation project, working for the governmental agency INDERENA from September 1971 to May 1973.

SUPPLY CHAIN

Barranquilla and Leticia, Colombia, together with Iquitos, Peru, are the three major centers exporting South American primates (Figure 1). Only 1,800 km (1,100 miles), a 3–4 hour trip by air, separates Barranquilla from Miami, the principal U.S. port of entry for neotropical fauna. Hence, most of the primates destined for export from the northern coastal area of Colombia leave via Barranquilla.* There are numerous sectors within this region from which monkeys are

*Colombians refer to the area of the northern Colombian coast and up to 400-km inland as "coast."

gathered (Figures 2 and 3). Yet, when received in Miami, the point of origin is known only as Barranquilla. In reality, the particular primate may have been captured 450 km or more from Barranquilla and, in some cases, in the Amazon.

One supply area is the Tiquisio region. I lived there in the town of Puerto Rico (Figure 2) for about a year and will describe the region and its primate trade. The physical terrain is characterized by low mountains between 300–1,000 m in altitude that begin the northern tip of the central cordillera of Colombia. This range joins the eastern and western cordillera to eventually form the single Andean chain in the southern part of the country.

Climatic conditions throughout the northern coast are characterized by constant tropical heat with variations related largely to altitudinal differences. Much of the region lies in the Magdalena River and Cauca River valleys and is affected by seasonal inundation. The rivers rise as much as 10 m during the rainy months, and the surrounding low-lying areas become floodplains and swamps, which are only passable by canoe or motorboat. The major rainy season is from the end of September through December, with a shorter period during April, May, and June.

Years ago, almost all of the Tiquisio area was covered by continuous evergreen primary tropical rain forest. Man, being the dominant exploiting species, moved into the area from the north to log, farm, and raise cattle. The Spaniards first settled Cocos, 15 miles south of Puerto Rico, over 150 years ago, to mine the area's gold deposits. Colorado, 19 km north of Puerto

85

FIGURE 1 Map of Colombia.

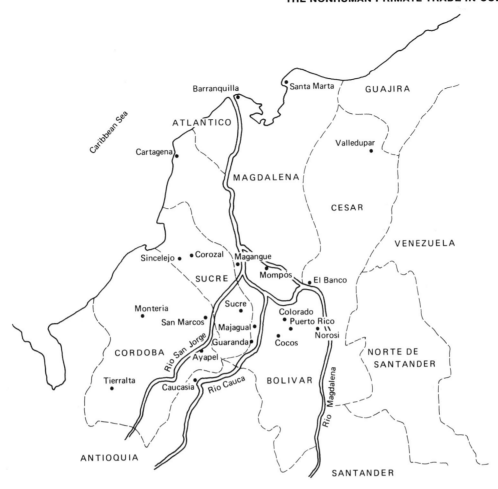

Map of the Colombian Coastal Region

0 km.   150 km.

FIGURE 2   Map of the Colombian coastal region.

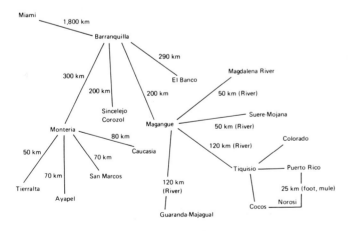

FIGURE 3   Supply chain for primates transported to Barranquilla.

Rico, was founded 80–100 years later. From these two villages, the populations expanded and dispersed to settle Puerto Rico 18–20 years ago.

Today most of the hillsides are scarred from slash and burn agricultural practices that utilize the lower lands for cattle pasture and steeper slopes for rice and corn crops. Yet, most areas are still pocketed with parts of primary forest next to well-developed secondary forest. The existing primary and secondary forest is able to support the following endemic platyrrhine species: *Alouatta seniculus, Aotus trivirgatus, Ateles belzebuth hybridus, Cebus albifrons, Saguinus leucopus,* and, in more restricted mountain localities, *Lagothrix lagotricha.*

Monkeys are not collected systematically in the Tiquisio area. In contrast to reports in the Iquitos and

Leticia regions (Middleton *et al.,* 1972), blowguns are not used, and neither box, drop, nor funnel trapping occurs. It is reported that *Cebus* are captured by mist-nets and fishnets. *Cebus* live in large social groupings of 15–20 and are infamous as crop pests. Campesinos (local inhabitants) are reported sometimes to surround the area where monkeys rest and sleep, to clear all trees until the monkeys are grouped in one or two trees, and to grab the desired monkeys when they attempt to escape. Common practices for collecting *Ateles* are to shoot an adult female carrying an infant or young juvenile and to collect the young animal when the mother falls to the ground. Very few people seem to eat monkey meat in the Tiquisio region, and thus a potential protein source for human consumption is wasted. A number of young juvenile *Ateles,* *Alouatta,* and *Saguinus* were caught by campesinos who were said to have climbed trees to catch them by hand.

All these species, with the exception of *Lagothrix,* can be found within a kilometer of Puerto Rico. Woolly monkeys inhabit the higher cooler areas further south, southeast, and southwest at a minimum distance of 20 km from the village. The woolly monkey of this region has a greyish cast over a black undercoat and is a distinct variety from the brownish Amazon race. Recently, Cooper and Hernández-Camacho (1976) reported that the woolly monkey, *Lagothrix lagotricha,* of this region represents the northernmost range of the genus and has been hunted in these sectors. This report contrasts with an earlier report by Thorington (1972) that states, "Woolly monkeys, *Lagothrix lagotricha,* are not found in the vicinity of Barranquilla."

Most campesinos are engaged in subsistence farming, and only a select few work land located close enough to the village to enable them to return daily. Since the only means of transportation to other parts of the country is by water, most campesinos come into Puerto Rico, as well as other villages, on Sundays to sell their crops. The crops are transported on Mondays by riverboats to Magangue and El Banco. Thus the so-called trappers are in reality campesinos who collect monkeys when time permits (when neither working their land nor fishing). Perhaps further supply fluctuations are due to the inaccessibility of the lowland regions when flooded during the rainy season. For example, during June and July 1972, the dry season was marked by severely dry weather resulting in the loss of rice and corn seedlings. It would be interesting to note if more time than usual is given to trapping when the next harvest months of September and October come. These factors contribute to the sometimes small numbers of monkeys leaving certain regions.

Handling of the captured animals varies little and is considered less than desirable or very poor. Most monkeys are kept in burlap sacks when transported to a village from the countryside. A small minority of the trappers transport monkeys in unscreened and unwired bamboo crates, which allow more circulation of air. The traveled distances may be 5–30 km, at times half a day's journey by horseback. The majority of animals suffer from exposure to the tropical sun, food deprivation during the journey, and the shock of being separated from the mother. Mortality and morbidity is common under this stress. Depending on the location, certain monkeys may be kept at the campesino's farm anywhere from 2 to 6 days before they are brought into the village to be sold.

Generally the trapper sells the monkey to a store merchant, who resells the monkey to persons who will eventually sell it to an animal dealer in Magangue. One animal dealer, who buys monkeys weekly from 10 to 12 trappers, travels directly to Barranquilla, bypassing Magangue. Also of interest is the fact that most monkeys leave the region on Mondays, the day that most weekly transportation goes upriver to Magangue. To further complicate matters, there may be one or two other resellings before a particular monkey finally reaches Magangue.

The riverboat journey lasts from morning to evening but may be shortened to 7–8 hours in a motorized dugout. In outboard motorboats, travel time is decreased to 4 hours, but few campesinos have access to this rapid means of transportation. Overall, the monkey is once again exposed to a drastic change in environment upon arrival at the deposit. At this point, depending on circumstances, most animals have passed through two to four hands.

Of the three present dealers in Magangue, one has been involved in the animal trade for 20 years, the second 15–16 years, and the third only 3–4 years. Therefore, these dealers have a preferred and sometimes exclusive clientele. There are no restrictions, however, and occasionally a trader will visit all three dealers several times before selling. Thus, the Magangue animal dealer needs only a small amount of capital to function, since he rarely ventures out to obtain the animals.

In addition to the Tiquisio area, the Magangue deposits receive monkeys that originate in the regions of the upper Cauca, San Jorge, and Magdalena rivers (Figures 2 and 3). *Cebus capucinus* (white-faced capuchin), *Saguinus oedipus* (cotton-top marmoset), and *Ateles fusciceps* (spider monkey) have a restricted geographical distribution along the west bank of the Magdalena River. These species are trapped in the vicinity of Sucre, Guaranada, and Majagual and at times sent to Magangue. Thus, Magangue supplies two species of capuchins, marmosets, and spider monkeys to Barranquilla.

It should be noted at this point that confusion exists in the use of regional names of certain species (Table 1). For example, *Cebus albifrons* is commonly called "cariblanca" in Tiquisio, yet in Barranquilla this name refers to *C. capucinus*. Similarly *Ateles* sp. are known as "mica," which closely resembles "mico," another name for *Cebus*. The worst example is "mico negro," which refers to *Lagothrix, C. capucinus,* and at times *Ateles fusciceps* (black spider monkey) coming from the Monteria region. As can be seen from this example, sectional records kept by persons referring to common name only must be scrutinized for such circumstances.

Conditions in the Magangue depositories are similar in that none have screened-in cages, the majority of which are made from wood and wire mesh. Two compounds have cages raised off the floor that allow most feces to drop through. One of these has several well-built, large, metal and wire mesh cages, 4 ft³, which hold up to a dozen *Aotus* each. The third has smaller cages lying on a cement floor that were rather unsanitary when observed. All are sheltered from direct sunlight and protected from rain, under a roof, and seem to be sufficiently ventilated. Food provided

for the monkeys ranges from well-matured and ripe "platano" (cooking banana), the local fruits when in season (papaya, guava, orange, and mango), and boiled milk. It is questionable to what extent the cages are cleaned after monkeys leave for Barranquilla, but it is doubtful that minimum suitable health conditions are met. One dealer has a small flamethrower with which he claims to sterilize cages, but I am skeptical of its frequency of usage.

Monkeys usually remain in the dealer's compound 1–6 days. Since most animals arrive Monday evening, the dealers try to transport them by bulk to Barranquilla on Tuesday, or whenever possible early in the week. Since Magangue is a large commercial riverport for this region of the Magdalena River, it is serviced by air, land, and water. There are daily DC-3 flights to Barranquilla (1 hour), regular bus departures to Barranquilla via Cartagena, covering 300 km of mostly asphalt highway (8 hours), and slower river transportation. One dealer regularly sends a shipment of *Aotus* in well-ventilated, double-screened, wooden boxes (the same type used to export monkeys overseas) on Tuesdays to Barranquilla via Avianca, the local airline. The other dealers use various modifications of the wire mesh wooden boxes, which are extremely small and confining for the animals, or the universal burlap bag.

Barranquilla has 20–25 commercial animal suppliers

TABLE 1 Common Neotropical Primate Names

Scientific Name	Local Common Name	English Common Name
Alouatta sp.	Cotudos, mono, mono rojo	Howler monkey
Aotus trivirgatus	Marta, mico de noche, martica, marteja	Owl, night monkey, Douroucouli
Ateles	Marimonda, mica, mela	Spider monkey
A. *belzebuth hybridus*	Marimonda gris, marimba	Long-haired spider
A. *fusciceps*	Marimonda negro, mico negro	Brown-headed spider
A. *geoffroyi*	Marimonda rojo	Black-handed spider
Cacajao rubicundus	Mico ingles	Red uakari
Callicebus torquatus	Zogui-zogui, mico socay	Dusky titi
C *moloch*		
Cebus	Mico, cariblanca, capuchino, monito	Capuchin, ringtail
C. *albifrons*	Cariblanca, mico bayo, mico maicero	Cinnamon ringtail
C. *capucinus*	Cariblanca, mico negro	White-faced ringtail
C. *apella*	Mico prego, mico azul	Weeper capuchin
		Hooded capuchin
Lagothrix lagotricha	Barrigudo, mico negro, mico churrusco	Woolly monkey
Pithecia monachus	Mico volador	Monk saki
Saimiri sciureus	Mico fraile, titi, vízcaino	Squirrel monkey
Cebuella pygmaea	Mico pielroja, titi de bosillo	Pygmy marmoset
Callimico goeldii	Leoncito	Goeldi's marmoset
Saguinus	Titi	Tamarin, marmoset
S. *leucopus*	Titi gris	White-footed marmoset
S. *mystax*	Bigote blanco	White-moustached marmoset
S. *nigricollis*	Bebeleche	White-lipped marmoset
S. *oedipus*	Titi blanco	Cotton-top marmoset

that negotiate primates and receive monkeys from regions of Magangue, El Banco, Sincelejo, and Monteria. These regions all have several localities similar to Magangue from which monkeys are trapped. Thus, the monkeys that arrive in Barranquilla may come from an area over 450 km away (Figures 2 and 3). I visited these facilities and questioned the dealers concerning their operations. It was evident that the majority of these suppliers had no firsthand knowledge of monkey trapping and did not know the length of time from capture to delivery at the compounds. Obviously certain information was withheld, which is understandable; but seemingly beyond the regions such as Magangue, the Barranquilla suppliers, without exception, do not know the exact geographic locations or sources of their monkeys.

Most of these animal dealers receive monkeys as well as other mammals, parrots, macaws, parakeets, snakes, and lizards daily. The animals come to Barranquilla via air and land, depending on the client and locality. Animals coming from Magangue have been discussed above. *C. capucinus* and *A. fusciceps* come by way of Monteria, San Marcos, and Valledupar. The cotton-top marmoset (*Saguinus oedipus*) comes from San Marcos and Monteria, originating in the areas of Tierra Alta, San Pedro (Antioquia), and Caucasio.

Conditions of these compounds are generally good or at least a step better than the sectional animal compounds of Magangue. Most have cement floors with drains, metal and wire mesh cages raised off the floor, well-ventilated and sheltered areas, and follow adequate husbandry practices. Crowding of monkeys was not evident, yet in one compound 10 *A. fusciceps* were kept in a large, walk-in, 5' × 3' × 3' cage together with two foxes. A second similar cage contained five infant *Tamandua*, seven *Sciurus granatensis*, two *Saguinus oedipus*, and one *Ateles fusciceps*. It was the exception to find screened walk-in units that had smaller screened and wired cages. Most suppliers feed their monkeys a basic diet of ripe platano and fruit, supplemented with rice and milk, bread, corn, or honey. Several said they added vitamins to the water, and one stated he added tetracycline.

The majority of the animal suppliers export monkeys to either the United States or other parts of the world, or both. Some specialize in dealing with supplying monkeys for biomedical and other scientific research. There are daily airline flights from Barranquilla to Miami. All monkeys are exported in double-screened, wooden crates, with sufficient ventilation, and have adequate waste disposal through the bottom of the box floor. During transport, most have water, and some suppliers even send bread with the animal as a food source for the longer journeys.

LEGISLATION AND CONTROL

INERENA, (*Instituto de Desarrollo de los Recursos Naturales Renovables,* i.e., Institute for Development of Renewable Natural Resources) is the Colombian governmental agency responsible for protecting the country's natural resources. Operating under the Ministry of Agriculture, it consists of several units. One, Parks and Wildlife, regulates use, hunting, management, transportation, and marketing of wildlife and their products. In addition, it concerns itself with biological studies, conservation, and repopulation of those species in danger of extinction.

INDERENA has a national headquarters in Bogota, with six regional offices throughout the country. The coastal region's headquarters, located in Barranquilla, has jurisdiction over the departments of Atlantico, Bolívar, Córdoba, Magdalena, and Sucre (Figure 2). Each region contains several sectional offices, which in turn control several local offices, one of which is Magangue. INDERENA has inspectors or rangers who are responsible for enforcing the various regulations in each sectional office. Magangue has three inspectors who issue licenses, collect fees, and patrol for illegal and contraband skins and animals that they have authority to confiscate.

All animals transported between any points in the country must have a "salvoconducto" (license to transport). Therefore, any monkey that is trapped must possess a salvoconducto if it is to be transported to Barranquilla. This salvoconducto contains data as to the type (scientific name never used), numbers, and monetary value of each monkey and may be obtained at the local office in Magangue. A small fee is collected, which supposedly becomes part of a "special INDERENA fund" (INDERENA, Resolution No. 564, 1970) for investigative wildlife studies. Another piece of legislation (INDERENA Acuerdo No. 18, 1970) states that commercial hunters should give 10 percent of their captured animals or the equivalent monetary value that includes capture, handling, and transportation to INDERENA for purposes of repopulation. (INDERENA, Resolution No. 36, 1971). This is rarely, if ever, done.

There are various problems of local administration that make it difficult, and at times impossible, for inspectors to enforce the laws. Three inspectors and one boat are far from enough to cover an area of jurisdiction that includes distances 100 km away. Secondly, cultural practices, political pressures, and at times graft inhibit any real regulatory efficiency.

Of particular interest are the regulations and laws pertaining to primates. Since 1941, 27 years before the existence of INDERENA, the Ministry of Agriculture had a law that prohibited hunting of deer and all classes of "pelo" (mammals) from March 1 to November 1 (Decreto No. 459 de 1951). In the past this law had not been enforced. Recently, the law has been utilized to enforce a closed trapping season for the hunting of primates. A salvoconducto, theoretically, can only be issued in Magangue if an animal supplier has obtained a permit from Bogota for the capture of a certain number and species of primates. Permits are issued only if the supplier can present a certificate of need from an established biomedical or research institution overseas that is certified by a Colombian consulate. For *Aotus*, permits may be issued that limit the number of individuals to be captured for all suppliers to 100 a month and a total of 700 during this closed season (INDERENA, Resolution No. 225, 1971). *S. oedipus* had been prohibited from being hunted throughout the year (Resolution No. 574, July 1969) because of its restricted range and fears of its extinction. Recent legislation (INDERENA, Resolution No. 568, 1972) amended the two previous resolutions in the northern coastal area, specifically to satisfy needs of the scientific and biomedical communities. The monthly capture limits for *A. trivirgatus*, *S. oedipus*, and *C. albifrons* are 100, 25, and 10, respectively, for each permit holder.

For a monkey to leave the country legally it must possess (1) a salvoconducto from INDERENA, which is a prerequisite for all of the following; (2) an exporting license from the controlling export agency, INCOMEX; (3) a certificate obtained from ICA (the agency controlling domestic and wild animal products), signed by a veterinarian or equivalent health official, stating that the animals are in good health and suitable for export; and (4) a clearance by customs when leaving the airport. In addition, there are several INDERENA inspectors at the airport responsible for checking animals destined for export to insure that all legal requirements have been satisfied.

One would think, with such regulations, that the flow of Colombian fauna would be well controlled. But, in reality, there is much illegal transport and exportation. INDERENA has inspectors that also check each animal compound in Barranquilla for irregularities. Yet I observed monkeys being transported illegally during the so-called closed season. Needless to say, this aspect of the primate trade is hard to document, yet there is substantial speculation as to the means of such activity. Monkeys may be shipped in crates labeled "raccoons," "anteaters," or other legally obtained animals for which a salvoconducto has been obtained. Thus the monkey can leave the country rather easily if controls are lax at any regulatory points.

Imported fish and wildlife entering the United States are regulated by provisions issued under the CFR (Code of Federal Regulations, 1970), Title 50—Wildlife and Fisheries, Part 17, known as "Conservation of Endangered Species and Other Fish and Wildlife." These laws, which became effective in June 1970, implement the Endangered Species Conservation Act of 1969 (83 Stat. 275), the Black Bass Act, and the Lacey Act.

At arrival in Miami the monkeys must meet the requirements of four governmental agencies: Public Health Service, Department of Agriculture, Department of the Interior, and Customs. A properly executed declaration for the importation of fish and wildlife (Form 3-177) must be submitted to Customs for collection of a 3.5 percent import tariff (Table 2). Information required includes common and scientific names, numbers, country of origin, whether or not the animal is on the endangered species list, and "whether or not subject to laws or regulations in any foreign country regarding its taking, transportation or sale" (Title 50, 17.4-b).

Inspectors of the Bureau of Sport Fisheries and Wildlife of the Department of the Interior examine the shipment to see that none are on the endangered species list. Listed neotropical primates include *Ateles geoffroyi frontatus*, *A. geoffroyi panamensis*, *Saimiri oerstedii*, *Brachyteles arachnoides*, *Chiroptes albinasus*, *Leontideus* sp., *Callimico goeldii*, and *Cacajao* sp. Only the latter two are found in Colombia's Amazon territory.

Presently, a rather shocking situation exists concerning Colombian exports. Monkeys enter during the closed season, yet action cannot be taken by U.S. authorities because the Bureau of Fisheries and Wildlife does not have all recent INDERENA laws on file. Title 50 further states that when wildlife is subject to foreign regulations, an export permit from an appropriate governmental agency, in this case INDERENA, must accompany the shipment.

The U.S. Public Health Service is concerned primarily with preventing the introduction of yellow fever from Latin America. South American monkeys must be in a mosquito-proof container upon arrival and have been maintained in mosquito-proof quarters for at least

TABLE 2 U.S. Customs Minimum Values for Live Colombian Monkeys (in U.S. Dollars)

Aotus	$ 5	*Lagothrix*	$40
Ateles belzebuth	$13	*Saimiri*	$ 5
Ateles fusciceps	$15	*Cebuella*	$ 5
Cebus albifrons	$10	*Saguinus nigricollis*	$ 5
Cebus capucinus	$13	*Saguinus oedipus*	$ 4

9 days prior to entry (Cooper, 1968). The certificate of health from ICA stating compliance with such regulations is sufficient evidence. These requirements are usually not followed. As mentioned earlier, animals are rarely kept in screened, mosquito-proof quarters in Barranquilla, and the ICA certificate received by Public Health cannot always be considered legitimate.

The Department of Agriculture inspects the containers for restricted fruits, plant matter, and insects.

MOVEMENT WITHIN COLOMBIA AND EXPORTATION

The number of monkeys transported within Colombia in 1970 is compiled by departments in Table 3. This table shows that 86.7 percent of the shipments were from the departments of Amazonas and Bolívar. These same two departments also supplied 56.2 percent of transported skins during 1970 (INDERENA, 1970). These figures demonstrate the concentration of faunal exploitation from the regions around Leticia and Barranquilla.*

During 1971, 2,381 monkeys were transported in the coastal region (Table 4). Of these, 65 percent were *Aotus*, 31 percent were *Cebus*, and 2 percent were *Ateles* sp. These data support earlier reports of the preponderance of night monkeys. The importance of Magangue, as compared to all other regions, is demonstrated by the fact that 97 percent of the *Aotus*, 57 percent of the *Cebus*, and 72 percent of the *Ateles* originated there.

The number of monkeys transported in the coastal region departments of Bolívar, Sucre, Córdoba, Atlantico, and Magdalena declined from 3,188 in 1970 (Table 3) to 2,275 in 1971 (Table 5).

Colombian records shown in Table 5 identified primates exported to the United States in 1970. These figures can be compared with the Fish and Wildlife Bureau's report (Paradiso and Fisher, 1972) that classified imports by country. INDERENA reported more than 10,839 primates were exported specifically to the United States, while the U.S. records show 12,988 Colombia primates imported in 1970 (Paradiso and Fisher, 1972). The records for individual species differ more than total values. Curiously, U.S. records show 833 *Saimiri* imported compared with Colombia's figure of 5,321. Data collected by different sources frequently differ, but it is unlikely that such a large number could have entered the United States without knowledge of the authorities. Data received from one very reliable

*The volume (11,734) transported nationally in 1970 according to Table 3 (INDERENA, 1971) compares favorably with the tabulation (9,142) from Table 5 (INDERENA, 1971).

TABLE 3 Number of Monkeys Transported in 1970 by Department

Department[a]	Number Monkeys	Percent
Amazonas	8,340	71.0
Bolívar	1,831	16.6
Sucre	470	4.0
Córdoba	457	3.9
Magdalena	430	3.8
Others[b]	206	0.7
TOTAL	11,734	100.0

SOURCE: INDERENA, Cuadro No. 40, 1971
[a] See Figure 1.
[b] Antioquia, Atlantico, Caquetá, Cesar, Meta, Putamayo, Santander, Norte de Santander.

animal dealer in Barranquilla shows that in 1970 he alone accounted for 80 *Ateles* sp., only seven short of the national total. A close look at national exports (Table 5) indicates that more *Aotus*, *Cebus*, and *Saguinus* were exported than transported. Table 6 illustrates the variation in governmental reports.

The majority of neotropical monkeys imported into the United States in 1970 came from Peru (60.9 percent) (USDI, 1971). Together with Colombian shipments (28.0 percent), these two countries represent a mammoth 88.9 percent of neotropical imports. Of the Colombian exports arriving in the United States, the following are percentages of the most frequently exported species compared with other Latin American countries (USDI, 1972): *Aotus*, 87.5; *Ateles*, 29.7; *Callicebus*, 52.8; *Cebus*, 47.4; *Lagothrix*, 21.8; and *Marmosets* 86.5. The tamarins, *S. oedipus* and *S. mystax*,

TABLE 4 Number of Monkeys Transported in the Coastal Region in 1971 by Section

Section[a]	*Aotus* (marta)	*Cebus* (mico cariblanca, mico bayo, mico maicero, mico)	*Ateles* (marimonda, marimba)
Magangue	1,544	416	43
El Banco	6	97	1
San Marcos	7	15	—
Monteria	12	31	14
Others	20	173	2
TOTAL	1,589	732	60

SOURCE: INDERENA. 1972 a, c—Note agreement in total primates transported in coastal region as reported by section (2381; INDERENA, 1971) and by species (2275; INDERENA.1972, Table 5).
[a] INDERENA offices that issue permits in the coastal region are located in these cities.

TABLE 5 Licensed Export and Transport of Colombian Primates

Scientific Name	Common Name	National Exported 1970 Total	To U.S.	To Others	National Transported 1970	Coastal Region[a] Exported 1971	Coastal Region[a] Transported 1971
Alouatta							
	Cotudo				102		
	Mono					4	17
Aotus		2,825					
	Marta		1,813	45		1,213	1,580
	Mico de noche		965	2	149		
Ateles		143					
	Marimba		87	28			
	Marimonda			28	251		
Cacajao	Mico ingles	20	20				
Callicebus	Zogui-zogui	44	44				
Cebus		266					
	Mico					807	290
	Mico bayo					260	50
	Mico maicero				54	127	326
	Mico cariblanca					11	
	Mico capuchinus		33	35		13	5
	Mico prego		198		108	74	
Lagothrix	Barrigudo	305	296	9	363	117	2
Pithecia	Mico volador	18		18			
Saimiri	Fraile	5,546	5,321	225	8,115	2,730	
Cebuella	Pielroja	165	157	8			
Callimico	Leoncito	8		8		1	
Saguinus	Titi	1,925	1,905	20		86	5
Subtotal		11,265	10,839	426	9,142	5,443	2,275
Undesignated		4,932					
TOTAL		16,197	10,839	426	9,142	5,443	2,275

SOURCE: National data adapted from INDERENA (1971); data for coastal region adapted from INDERENA (1972a, b, c, d).

[a] Coastal region includes the departments of Atlantico, Bolívar, Córdoba, Magdalena, and Sucre.

account for almost 90 percent of the Callitrichidae. This family roughly corresponds to one-third of all Colombian primates entering the United States for that year. Approximately 60 percent of these species im-

TABLE 6 Capture and Export of Colombian Monkeys

Number of Monkeys	1969	1970
Registered captures	18,572[a]	11,496[c]–20,882[a]
Exported	18,522[b]	14,519[b]–16,197[c]

[a] INDERENA, 1972.
[b] INDERENA, 1972.
[c] INDERENA, 1971.

ported into the United States were utilized for research purposes. This is a slight increase over 1969 (Table 7).

Aotus and *Saimiri*, together with *Cebus albifrons* and *C. apella*, were also utilized heavily. As can be seen, the number of squirrel monkeys not used for research purposes (all years) is significant. The squirrel monkeys, 57 percent of New World imports for 1970, represent a numerical drop from previous years. Caution should be exercised, however, since the number of unspecified exports for this period as reported in Colombian records is 4,524. Flow into the pet market accounts for almost all of these nonresearch monkeys. Yet, imports dropped 46 percent from 1969.

Most striking and unfortunate is that the number of *Cebus* used outside of research remained alarmingly high, close to 6,000 animals (approximately 85 percent). For all species the following percentages were

TABLE 7   Importation and Research Use of Neotropical Primates, 1968–1970

Species	Common Name	Imported 1968[a]	Used for Research 1968[b]	Imported 1969[c]	Used for Research 1969[d]	Imported 1970[e]	Used for Research 1970[f]
Alouatta	Howler monkey	13		36		1	
seniculus	Red howler	16		12		12	
palliata	Mantled howler	101		93		88	
Aotus trivirgatus	Night monkey Douroucouli Owl monkey	4,087	5,470	5,312	3,972	4,327	4,519
Ateles	Spider	280	981	77	619	14	228
belzebuth	Long-haired spider	142		31		87	
fusciceps	Brown-headed spider	34	19				
geoffroyi	Black-handed spider	1,105		1,548		1,870	
paniscus	Black spider	712		984		587	
Brachyteles							
arachnoides	Woolly spider	1		38			
Cacajao melanocephalus	Black-headed uakari	4					
rubicundus	Red uakari	127		69		14	
Callicebus	Titi	141	17	116	159	120	10
moloch	Dusky titi	51		43		40	
Cebus	Capuchin	106	1,740	99	1,058	101	916
albifrons	White-fronted	4,913		4,743		3,168	
apella	Black-capped	784		1,024		847	
capucinus	White-throated	1,768		1,574		1,762	
nigrivittatus	Weeper	106		78		36	
Chiropotes satanas	Black saki	4		6		14	
Lagothrix lagotricha	Humboldt's woolly	2,902	1	3,311		2,241	
Pithecia monachus	Monk saki	162		24		66	
pithecia	Pole-headed saki	10		2		2	
Saimiri sciureus	Squirrel	45,014	20,616	47,096	8,429	26,124	6,807
Not designated			114		265		
SUBTOTALS (Cebidae)		62,583	28,823	66,333	14,343	41,522	12,480
Callimico goeldii	Goeldi's marmoset	83		43		9	
Callithrix	Marmoset	22		1		44	
aurita	White-eared	31		446		126	
argentata	Black-tailed	57		40		22	
jacchus	Common	125		52		1	
Cebuella pygmaea	Pygmy marmoset	197		639		192	
Leontopithecus rosalia	Golden lion tamarin	50		149		150	
Saguinus	Tamarin	89				1	
graellsi	Rio napo	3		6			
illigeri	Redmantled	289		197			
labiatus	Red-bellied	92		9			
leucopus	White-footed	33					
nigricollis	Black and red	3,519		1,564		332	
oedipus	Cotton top	3,098		3,752		2,068	
geoffroyi	Geoffroy's			6		9	
midas	Yellow-handed			1			
tamarin	Negro			32			
mystax	White-moustached						1,780
SUBTOTAL (Callitrichidae)		7,688	3,858	6,537	2,415	4,734	2,804
TOTAL		70,271	32,799	72,870	16,758	46,256	15,284

[a] Jones, 1970.
[b] ILAR, 1969.
[c] Jones and Paradiso, 1970.
[d] USDI, 1971.
[e] Paradiso and Fischer, 1972.
[f] ILAR, 1971b.

TABLE 8 Number of Animals Imported into the United States

	1968[a]	1969[b]	1970
Mammals	129,520	122,991	101,302[c]
Total primates	113,714	108,974	85,151[c]
			(78,375)[d]
New World primates	70,271	72,870	46,256[d]
Old World primates	43,443	36,099	31,280[d]

[a] Jones, 1970.
[b] Jones and Paradiso, 1970.
[c] USDI, 1971.
[d] Paradiso and Fischer, 1972.

obtained by dividing the number of research monkeys by the total number of imports: 47 percent (1968), 23 percent (1969), and 33 percent (1970). It is depressing to see the numbers that apparently reached private hands as pets, especially during 1969. To those concerned with the exploitation of South American primates, it is important to note that the import volume in 1968 and 1969 dropped in 1970. The numbers of New World primates utilized for pet purposes in 1970 were nearly 20,000 *Saimiri*, 5,000 *Cebus*, and 2,200 *Lagothrix* (none reported to be used for research).

Both New World and Old World primates accounted for 87.8, 88.6, and 84.4 percent of all mammals entering the United States in 1968, 1969, and 1970, respectively (Table 8). Further breakdown indicates that New World monkeys were imported at consistently higher levels than Old World monkeys for the same years—61.8, 66.9, and 59.7 percent.* At the Miami port of entry, 57,921 primates entered in 1970, with neotropical primates probably comprising at least 60–70 percent.

Now that a general picture of the supply chain has been presented, monetary returns at each step of the chain can be examined (Table 9). This list, as compiled by the author during the past year, is believed to be relatively accurate. Prices quoted are what the dealer pays for the animal locally, in Magangue and Barranquilla. Also included are the export prices quoted from dealers for monkeys delivered at Miami and a representative sample of pet store prices. Undoubtedly, transport costs of $0.44 per lb for exportation are significant to the exporter, yet the margin of profit at times is believed to be close to 100 percent. One pet store

*Note in Table 8 that the 1970 import volume reported in USDI (1971) is 6,776 more than that reported in Paradiso and Fisher (1972).

visited in Miami wanted a price of $175 for an infant woolly monkey. It is distressing to see how these profits are disproportionately inflated in the United States when Colombia, which has permanently lost this natural resource, receives little of the economic gain. The campesino who traps the animal also reaps little of this eventual monetary gain, yet in local economic terms profits well. Normal weekly wages in the Tiquisio area are 30 pesos (U.S. $1.50), and the sale of one woolly monkey amounts to 2 weeks' work. It is evident that very few individual trappers and local sectional dealers have taken the main profits.

The dollar volume of Colombian primate export to the United States was the following:

1969	*1970*
1,671,400 pesos	2,297,000 pesos
U.S. $79,543	U.S. $109,381
	($73,951)*

Economically, primate exportation rated below alligator skins, live snakes, peccary hides, and live cats. The economic value of the monkey trade represents a very small percentage (3 percent) of Colombia's total faunal export credit.

CONSERVATION

As can be deduced from the facts presented in the previous pages, various pressures exist that are affecting the nonhuman primate population. As a result of commercial hunting and trapping for the pet and scientific community and competition with expanding human populations, such populations may be reduced in Colombia as well as throughout the world. In broader terms, the expanding human population results in nonhuman primate habitat deterioration and/or destruction, reduction in the numbers of monkeys by eliminating them as agricultural pests, and utilizing them as a food source.

Past reports have varied in claimed mortality rates due to trapping and transportation along the supply chain. Most of these reports lack documented evidence in South America, and statements such as the following do little to promote accurate information:

Conservative estimates indicate a loss of 50% during the period from the capture of the animal until its final sale. (Avila-Pires, 1968)

*Tabulated from INDERENA, 1971, Cuadro No. 3; other figures from INDERENA, 1972, Cuadro No. 38, using 21 pesos = U.S. $1.00 as conversion factor.

TABLE 9 Price Chain per Colombian Primate

Type of Primate	Price to Local Buyer		Price at Magangue		Price at Barranquilla		Barranquilla Export Price		Price at U.S. Pet Stores	
	Pesos[a]	U.S.$[a]	Pesos	U.S.$	Pesos	U.S.$	Pesos	U.S.$	Pesos	U.S.$
Aotus	40–50	2–2.5	50	2.5	70	3	95–150	4.5–7	630–840	30–40
Ateles	90–150	4–7	150	7	200	9.5	273–420	13–20	1,050–1,680	50–80
Alouatta	50–80	2.5–4	80–100	4–5	100–130	5–6	126–210	6–10	—	—
Cebus										
alibifrons	60–80	3–4	80	4	120	6	210–263	10–13	1,575–1,689	75–80
capucinus	—	—	—	—	—	—	263–420	—	1,689–2,100	—
Saguinus oedipus										
and leucopus	20–40	0.9–2	40–45	1.9–2	50	2.5	95–170	4.5–8	735	35
Lagothrix	200–300	10–14	300–500	14–24	300–500	14–24	950	45	3,675–4,200	175–200
Saimiri	Not in coastal region						168–315	8–15	735–840	35–40

[a] Conversion Rate = 21 Colombian pesos/U.S.$. U.S. figures are approximated.

Thorington (1972), on the other hand, has made an accurate and honest statement regarding such speculations:

> . . . we are, however, unaware of the number of animals trapped in South America, those which die in trappers and dealers compounds or en route and the causes of mortality. . . . (p. 18)

Emphasis must be placed on the lack of clear, documentary material for determining these mortality and morbidity percentages. One reliable dealer in Barranquilla reported that *Aotus* had a mortality rate between 45–50 percent in his compound until he restricted his source of supply solely to Magangue, resulting in a mortality drop to 5–8 percent.

The second major consideration of trapping pressures is the quantity of primates extracted from the various localities. Some species, such as *Aotus* and *Saimiri sciureus*, are heavily harvested yet probably are not near danger of extinction because of wide distribution and abundance. On the other hand, certain populations and species, even though trapped to a more moderate degree, may be under greater stress due to their restricted distributions. As mentioned earlier, the cotton-top marmosets are only found west of the Magdalena and north of the San Jorge rivers, and the woolly monkeys in this region are geographically restricted to two small mountainous zones (Hernández-Camacho and Cooper, 1976). Similar numbers of *Ateles* and *Lagothrix lagotricha* are exported to the United States, even though woolly monkeys have a very restricted range, in contrast to *Ateles*, which are widely distributed within Colombia. Therefore, the numbers of animals collected in a given locality do not necessarily reflect the numbers of animals in the natural habitat.

One of the greatest threats to primate populations is the alteration of environment, resulting in a reduction of suitable habitats for these species. In northern Colombia, lumbering, cutting roads, and slash and burn agricultural practices result in habitat destruction. The rate of deforestation is increasing exponentially due to Colombian demographic expansion. The species of *Alouatta*, *Aotus*, *Ateles*, *Cebus*, and *Saguinus* have all been observed to inhabit various types of secondary and primary forest throughout Tiquisio and the Magdalena River valley. Frequently, populations can be found in close proximity to farmland. Campesinos generally use such land for 2 or 3 years, then allow it to lie fallow for several years, during which time native vegetation regenerates rapidly.

Reports in Tiquisio mention that *C. albifrons* and *S. leucopus* raid the crops of many campesinos. This suggests that small numbers of these monkeys are eradicated as agricultural pests. A few monkeys are hunted for food. Since only a select number of natives possess firearms, and ammunition is expensive and in short supply, these losses are minimal.

It can be seen from this discussion that in certain areas greater land use does not necessarily result in complete habitat destruction. Although the following statement may be true for parts of South America, it is not applicable to northern Colombia nor Barranquilla and probably reflects the situation in an emotional rather than realistic manner:

> All nonhuman primates within the hunting ranges of Iquitos, Leticia, and Barranquilla are threatened and severely decimated. (Harrisson, 1971, p. 7)

It is hoped that such statements would be subjected to serious scrutiny to obviate any doubts of authenticity.

With any discussion of the primate trade, one should

incorporate topics that need further attention. Past reports have discussed and proposed in detail needed research similar in tone to the statement by Southwick *et al.* (1970) that:

It becomes increasingly incumbent upon mankind in general and the scientific community in particular to undertake more vigorous research and conservation programs to protect endangered (primate) species. (p. 1051)

It is believed the scientific community has been well saturated with these suggestions and demands. With respect to Colombia and that of all South America, very little has been done to promote ecological investigations of different species, to determine the abundance and natural distributions of these populations, to analyze the impact of the supply trade on remaining populations, to accumulate data to determine what level of harvest can be sustained, and to identify present irrational uses of this resource.

The Colombian legislation that presently attempts to regulate and manage its primate resources has already been mentioned. INDERENA has far from an ubiquitous system of control, but many measures, even though not strictly enforced, do contribute to a reduction in primate exploitation. Unfortunately, there are certain anomalies. For instance, why does the open trapping season exist in November, December, January, and February? These winter months in the Northern Hemisphere correspond to the most stressful period of the year because of the drastic climatic change. Considering that the United States receives the majority of Colombian primate exports, it seems reasonable to question this regulation. Since this decree (No. 459) was issued in 1941 with no known biological basis, it seems imperative that it be reviewed and possibly altered for better management purposes.

In terms of action and control in the near future, INDERENA has considered legislating the following (Hernández-Camacho and Cooper, 1976):

• Setting aside natural reserves to protect certain threatened species, subspecies, or populations.
• Limiting exploitation of its remaining natural primate resources specifically to biological and biomedical research.

Also, a feasibility and cost-analysis study of breeding primates in captivity through programs of "zoocriaderos" (animal farming compounds) is under consideration. Hence, at the local level there is genuine interest and activity to varying degrees.

Besides national regulations, there should be a more acute system of coordinated action between countries such as Colombia and the United States. Previously mentioned is the absence of Colombian laws on record with the U.S. Department of the Interior. Other possible measures include greater return of economic profits from the pet market in the United States back into the Colombian economy. Perhaps the U.S. Government should drastically reduce or completely eliminate the primate pet market altogether. But a necessary note of caution must be interjected. Such restrictive measures might stimulate undesirable and illegal activity that already exists in Colombia, or even worse inhibit legitimate operations.

Internationally, such organizations as the World Wildlife Fund (WWF) and the International Union for the Conservation of Nature and Natural Resources (IUCN) have action committees to propagate conservation activities in all fields, including neotropical primates. Basically, the practice is to set aside preserves for protection and management of critical habitats and species. It must be stressed that such advice and encouragement is helpful, but only purposeful when coupled with funds. The Fourth International Primatological Society (IPS), meeting in Portland, Oregon, on July 14–18, 1972, drafted an appeal for conservation of nonhuman primates. It is hoped that this will be a productive beginning for stronger international cooperation.

ACKNOWLEDGMENTS

I would like to express my gratitude to the Peace Corps for enabling me to live in Colombia during 1971 to 1973, as well as my supporting agency, INDERENA. Particularly, I would like to mention the assistance given by Dr. Jorge Hernández-Camacho, chief of Wildlife, and Dr. R. W. Cooper, of the Peace Corps Conservation Program. ILAR provided the funds for travel to the conference. In Miami, Mr. John Thomas, the USDI Sport Fisheries and Wildlife inspector, was extremely helpful and cooperative. Finally, the moral support and encouragement of my wife, S. Huffman, in Colombia, was invaluable, as well as the constructive recommendations by Nancy A. Muckenhirn.

REFERENCES

Avila-Pires, F. D. de. 1968. Some problems concerning primates. Pages 132–141 *in* Proceedings of the Latin American Conference on the Conservation of Renewable Natural Resources, San Carlos de Bariloche, Argentina, 27 March–2 April, 1968. International Union for Conservation of Nature and Natural Resources (IUCN), Morges, Switzerland.

Code of Federal Regulations. 1970. Conservation of endangered species and other fish or wildlife, Part 17, Title 50. Federal Register, vol. 35, no. 106.

Cooper, R. W. 1968. Squirrel monkey taxonomy and supply. Pages 1–29 *in* L. A. Rosenblum and R. W. Cooper, eds. The squirrel monkey. Academic Press, New York.

Cooper, R. W., and J. Hernández-Camacho. 1975. A current appraisal of Colombia's primate resources. Pages 37–66 *in* G. Bermant and D. G. Lindburg, eds. Primate utilization and conservation. John Wiley and Sons, Inc., New York.

Harrisson, B. 1971. Primates in medicine. Vol. 5. Conservation of nonhuman primates in 1970. S. Karger, Basel. 98 pp.

Hernández-Camacho, J., and R. W. Cooper. 1976. The nonhuman primates of Colombia. This volume.

ILAR. 1969. Annual survey of animals used for research purposes during calendar year 1968. ILAR News 13(1):i–xi.

ILAR. 1970. Annual survey of animals used for research purposes during calendar year 1969. ILAR. News 14(1):i–v.

ILAR. 1971a. Annual survey of animals used for research purposes during calendar year 1970. ILAR News 15(1):i–ix.

ILAR. 1971b. Survey of nonhuman primates being maintained as of January 1971, prepared by R. W. Thorington, Jr. ILAR News 15(1):7–10.

INDERENA. 1970. Cuadro No. 37.

INDERENA. 1971. Anuario de la fauna silvestre en Colombia en 1970. Oficina de Planeacion, Cuadro No. 3. 87 pp. Mimeographed report.

INDERENA. 1972a. Movilizacion de los recursos faunisticos. Regional Costa Atlantico, Informe Estadistico, Primer Semestre, 1971, Enero–Junio, Barranquilla. Mimeographed report.

INDERENA. 1972b. Exportacion de los productos faunistico. Regional Costa Atlantico, Informe Estadistico, Primer Semestre, 1971, Enero–Junio, Barranquilla. Mimeographed report.

INDERENA. 1972c. Movilizacion de los recursos faunisticos. Regional Costa Atlantico, Informe Estadistico, Segundo Semestre, 1971, Julio–Diciembre, Barranquilla. Mimeographed report.

INDERENA. 1972d. Exportacion de los productos faunistico. Regional Costa Atlantico, Informe Estadistico, Segundo Semestre, 1971, Julio–Diciembre, Barranquilla. Mimeographed report.

INDERENA. 1972e. Los recursos naturales renovables del Pais en Cifras. Documento Provisional, Bogota.

Jones, C. 1970. Mammals imported into the United States in 1968. Special scientific report. USDI WL-137. U.S. Government Printing Office, Washington, D.C. 30 pp.

Jones, C., and J. L. Paradiso. 1970. Mammals imported into the United States in 1969. Special scientific report. USDI WL-147. U.S. Government Printing Office, Washington, D.C. 33 pp.

Middleton, C. C., A. F. Moreland, and R. W. Cooper. 1972. Problems of New World primate supply. Lab. Primate Newsl. 2(2):10–17.

Paradiso, J. L., and R. D. Fisher. 1972. Mammals imported into the United States in 1970. Special scientific report. USDI WL-161. U.S. Government Printing Office, Washington, D.C. 62 pp.

Roth, W. T. 1965. Editorial. Primates from trap to test tube. Lab. Anim. Care 15(4):243–246.

Roth, W. T. 1969a. Primate supply today. Pages 206–210 *in* Proceedings of the 2nd International Congress of Primatology, Atlanta, Georgia, 1968. Vol. 2. S. Karger, Basel.

Roth, W. T. 1969b. Supplying wild primates to the laboratory. Pages 1–16 *in* W. I. B. Beveridge, ed. Primates in medicine, vol. 2. S. Karger, Basel.

Southwick, C. H., M. R. Siddiqi, and M. F. Siddiqi. 1970. Primate populations and biomedical research. Science 170:1051–1054.

Thorington, R. W., Jr. 1972. Importation, breeding, and mortality of New World primates in the United States. Int. Zoo. Yearb. 12:18–23.

USDI (U.S. Department of the Interior), Bureau of Sport Fisheries and Wildlife. 1971. Wildlife imported into the United States in 1970. USDI WL-495. U.S. Government Printing Office, Washington, D.C.

ADDENDUM TO THE NONHUMAN PRIMATE TRADE IN COLOMBIA

Nancy A. Muckenhirn

The primate trade in Colombia between 1968 and 1970 was described by K. Green based upon figures compiled by Instituto de Desarrollo de los Recursos Naturales Renovables and the U.S. Department of the Interior (USDI). More recent estimates have amplified our understanding of the trade described in the previous paper. Recent statistics have shown that the proportion of the primate trade used in research in the United States has been underestimated, sizable discrepancies have existed among most estimates of import volumes when they are collected by different methods or agencies, and there has been an overall decline in primate imports during the past 6 years.

The recent USDI estimates of primate imports from 1971 and 1972 are presented in Table 1. Colombia supplied more than half of 12 species in at least one of these years, including *Alouatta seniculus*, *A. palliata*, *Aotus trivirgatus*, *Ateles belzebuth*, *A. fusciceps*, *Callicebus moloch*, *Cebus albifrons*, *C. capucinus*, *C. nigrivittatus*, *Cebuella pygmaea*, *Saguinus nigricollis*, and *S. oedipus*. Several hundred *Saguinus mystax* and a few *Callithrix chrysoleuca* were also recorded from Colombia, although these species do not originate there. The species exported from Colombia in the largest numbers were squirrel monkeys (5,000–6,000), night monkeys (3,000), *Saguinus oedipus* (2,000–2,300), *Cebus albifrons* (1,300–2,200), and *C. capucinus* (1,000).

The estimates cited by Green for neotropical primates used in research were derived from annual surveys by the Institute of Laboratory Animal Resources (ILAR). These values were compiled largely from estimates reported by commercial suppliers of their sales to research users. Surveys of primate inventories in research institutions were made in 1971 and 1973 (Thorington, 1971; Muckenhirn, 1975). The inventories are compared to import figures for the corresponding years in Table 1 and are not strictly comparable to the estimates of research use in the earlier ILAR surveys. Muckenhirn (1975) concluded that the research use of primates has been underestimated as a result of incomplete sampling of research institutions. In addition, the past estimates of the volume of the pet trade appear to have been unrealistically high, because the estimates of the pet trade have been derived by subtracting the low estimates for research use from total imports and they have not included estimates of sizable dealers' losses during quarantine.

The trends in imports of Latin American primates during the past decade are presented in Table 2. The proportion of neotropical primates among total imports was 45 percent during the first 3 years of the decade, remained between 53 and 63 percent during the next 6 years, and dropped to 23 percent in 1973. The curtailment by Colombia in 1974 reduced the export volume from that country to 10 percent of the peak 1968 level. This decrease in Latin American imports was largely responsible for the overall decline to 30 percent of their peak 1968 level. The establishment of a permit system by INDERENA for exporting primates to research institutions has effectively eliminated the pet trade in primates from Colombia. The establishment of export quotas may be expected to

TABLE 1 Imports and Inventories of Neotropical Primates in the United States

Species	1971 Imports	1971 U.S. Research Inventory	1972 Imports	1973 U.S. Research Inventory
Alouatta caraya	18	—	4	—
A. seniculus	11	—	7	—
A. palliata	47	—	33	—
Aotus trivirgatus	3728	1061	3533	1316
Ateles belzebuth	41	—	34	—
A. fusciceps	129	129	70	—
A. geoffroyi	1617	3	1841	30
A. paniscus	82	9	122	4
Ateles sp.	20	11	3	64
Callicebus moloch	40	34	24	39
Callicebus sp.	134	4	42	—
Cebus albifrons	2221	255	2776	261
C. apella	2036	177	1975	511
C. capucinus	1133	11	1209	5
C. nigrivittatus	33	—	—	—
Cebus sp.	196	14	103	50
Chiropotes satanas	—	—	6	—
Lagothrix lagotricha	2226	22	2125	29
Pithecia monachus	83	—	30	—
P. pithecia	1	—	—	—
Saimiri sciureus	29,879	3941	25,297	4358
Subtotal Cebidae	43,675	5542	39,234	6667
Callimico goeldi	—	13	4	13
Callithrix aurita	—	—	48	—
C. chrysoleuca	—	—	19	—
C. goeffroyi	24	—	—	—
C. jacchus	3	—	9	186
C. penicillata	26	—	—	4
Callithrix sp.	—	—	10	4
Cebuella pygmaea	166	44	111	2
Saguinus geoffroyi	12	—	11	2
S. fuscicollis	—	317	—	1199
S. illigeri	293	—	50	5
S. mystax	863	462	1064	717
S. nigricollis	1787	1359	1933	441
S. oedipus	2374	390	2419	614
S. tamarin	—	1	—	—
Marmoset	3	—	63	151
Subtotal Callitrichidae	5552	2585	5746	3341
TOTAL	49,227	8127	44,980	10,008

SOURCES: Thorington (1971), Muckenhirn (1975), Clapp and Paradiso (1973), USDI (in prep.).

TABLE 2 Neotropical Primate Imports to the United States

Year	Peru	Colombia	Total Latin America	Total Primates
1964	36,847	6,841	45,630	102,080
1965	33,634	9,123	43,628	96,112
1966	37,384	9,491	47,266	103,859
1967	39,600	13,879	55,011	104,346
1968	53,773	24,105	80,124	126,857
1969	45,980	17,563	64,925	105,719
1970	32,729	16,826	53,000	90,743
1971	31,550	15,910	49,879	79,846
1972	27,288	16,124	45,414	75,784
1973	22,669	6,444	30,871	69,548
1974	2,251	2,313	10,869	46,581

SOURCE: U.S. Department of Commerce, 1965, 1966–1967, 1968–1973, 1974, 1975.

continue to limit the numbers of wild-trapped primates available in the future from this traditional supplier country.

REFERENCES

Clapp, R. B., and J. L. Paradiso. 1973. Mammals imported into the United States in 1971. Special scientific report. USDI WL-171. U.S. Government Printing Office, Washington, D.C.

Muckenhirn, N. A. 1975. Supporting data. Pages 11–122 in Nonhuman primates: usage and availability for biomedical programs. Committee on Conservation of Nonhuman Primates, Institute of Laboratory Animal Resources, National Academy of Sciences, Washington, D.C.

Thorington, R. W., Jr. 1971. Summary of nonhuman primates being maintained on 1 January 1971. ILAR News 15:7–10.

U.S. Department of Commerce, Bureau of the Census. 1965. U.S. imports, tariff schedules annotated by country, 1964 annual. Washington, D.C.

U.S. Department of Commerce, Bureau of the Census. 1966–1967. U.S. imports, TSUSA [tariff schedule of the U.S., annotated], commodity by country 1965 and 1966 annual. Washington, D.C.

U.S. Department of Commerce, Bureau of the Census, 1968–1973. U.S. imports for consumption and general imports. TSUSA [tariff schedule of the U.S., annotated], commodity and country. USDC FT-246, 1967–1972 annual. Washington, D.C.

U.S. Department of Commerce, Bureau of the Census. 1974. U.S. imports for consumption TSUSA [tariff schedule of the U.S., annotated], commodity and country of origin. USDC IM-146, December 1973. Washington, D.C.

U.S. Department of Commerce, Bureau of the Census. 1975. U.S. imports for consumption. TSUSA [tariff schedule of the U.S., annotated] schedule by TSUSA number by unit control of origin. USDC IM-146, 1974 annual. Washington, D.C.

USDI (U.S. Department of the Interior). In preparation. Mammals imported into the United States in 1972.

THE POPULATION AND CONSERVATION OF HOWLER MONKEYS IN VENEZUELA AND TRINIDAD

Melvin Neville

INTRODUCTION

The demographics of a population of howler monkeys on Hato Masaguaral, a cattle ranch in Venezuela's Guárico State, is examined with regard to the practical difficulties in censusing. Extended work in limited areas with repeated age–sex counts on each troop is advocated. Census data from work in 1969–1970 and 1972 on Hato Masaguaral are analyzed to show the need for a large sample size in estimating such parameters as sex ratio and to provide an idea of the population characteristics of howler monkeys in a protected llanos area.

Hill (1962) lists five subspecies of *Alouatta seniculus* for Venezuela and Trinidad. *A. s. seniculus* occurs on the western border with Colombia. *A. s. stramineus* and *A. s. macconelli* are found south of the Orinoco. *A. s. arctoidea* in the northern part of Venezuela is the nearest mainland neighbor of *A. s. insulanus* on Trinidad. The dominant capuchin in most of Venezuela is *C. nigrivittatus bruneus* (see Hershkovitz, 1949; *C. griseus grisens* of Hill, 1960). Thus, *C. nigrivittatus* intervenes between the Trinidad capuchin, *C. albifrons trinitatis,* and the nearest mainland subspecies of *C. albifrons.* Hershkovitz (1972) states that "the Trinidad capuchin was, almost certainly, introduced by man."

We know little about the total extent of the distribution of the red howler monkey, *A. seniculus,* in Trinidad and Venezuela. The extension of the population data that I obtained at my study sites becomes increasingly speculative when extrapolated to larger areas. For this reason, I will avoid estimating numbers of monkeys that may remain in other areas. I close instead with a discussion of the possibilities for conservation of howlers and capuchins in Trinidad and Venezuela.

STUDY SITES, METHODOLOGY, AND TECHNICAL COMMENTS

The actual fieldwork for the demographic portions of this chapter consisted of 51.5 h of observations of howling monkeys in Bush Bush Forest, Trinidad, during the summer of 1968; 603 h of observation of howlers (out of ca. 893.5 field hours) at Hato Masaguaral in Venezuela in 1969–1970; and 58 h of observations on *Alouatta* during June and July of 1972 at Hato Masaguaral (out of 82.5 field hours). Additional field time was spent at Camatagua Reservoir in Venezuela during the summer of 1972.

Hato Masaguaral (Figures 1 and 2), a cattle ranch owned by Tomás Blohm, is situated approximately 50 km south of Calabozo in southern Guárico State. The ranch provides a sample of primate populations in a protected or unutilized area in the llanos north of the Orinoco and will form the basis for my discussion of demography. The Hato has extensive populations of howlers (*A. s. arctoidea*) and limited numbers of capuchins (*C. nigrivittatus*) (Neville, 1972). The Camatagua Reservoir, formed in 1969 near Camatagua in southeastern Aragua State, regulates the flow of the

FIGURE 1   Venezuela: 1–7, national parks on the UN list (IUCN, 1967); 8–9, howler study sites. 1. Henry Pittier (Rancho Grande). 2. Guatopo. 3. Sierra Nevada de Mérida. 4. El Avila. 5. Yurubi. 6. Yacumbo. 7. Canaima. 8. Camatagua Reservoir. 9. Hato Masaguaral.

Guárico River and supplies water to Caracas. Its watershed has extensive howler and limited capuchin populations.

The basic approach to censusing was to record directions of dawn howls, track down the troops, and make counts. However, *A. seniculus* troops do not always participate in the dawn chorus (a variability also reported by Chivers, 1969, for *A. palliata* on Barro Colorado Island). Chorusing frequencies range from 96 percent (22 of 23 mornings for the WHT troop from Bush Bush) to 18 percent (4 of 22 mornings for Troop 1 of Western Forest, Hato Masaguaral). One of the possible factors influencing the variability is density in a limited region. However, even some of the most densely packed troops occasionally failed to participate in the dawn howl.

Walking through an area and counting animals, one may miss entire troops. Howlers may remain unnoticed in a small tree unless an animal happens to move or vocalize. Vocalizations were a fairly common response to humans in Hato Masaguaral, where most of the troops were incompletely habituated and unhunted. Where hunting occurs, however, the animals are silent and will remain motionless unless able to flee. In any given area, I believe that an accurate census can only be obtained by repeating counts of each resident troop at least twice. If a troop is censused only once, there is a high probability of error due to hidden or strayed monkeys. Figure 3 shows that the 1969–1970 counts for Middle and Eastern Forests in Hato Masaguaral continued to increase as more time

was invested and I became better acquainted with the area. All but two troops were counted twice in the 1972 Hato census. For 5 of the 17 troops that were counted more than once, the first count was incomplete.

Correct ascertainment of sex and age also profits by repeated counting. Howlers also are notoriously difficult to maintain in zoos, hence I was not able to obtain a mental image of size, shape, and behavioral changes correlating with age classes before going into the field. The correlation of field categories to a chronological scale had to be made in retrospect and is therefore improving with continued work.

The basic criteria for age assignment in this report were shape of the female genitalia, the male head and throat, and size. Animals assigned as infants were approximately 0 to 10 or 12 months of age. The juvenile stage appeared to last from ca. 1 yr of age to 2.5 yr, the criterion for the end point being a change in the female genitalia with the clitoris losing prominence relative to the labia. The size achieved by females of this stage was used to define the lower size limit for small subadult males. The subadult female stage appeared to last approximately 1 yr, until the female achieved essentially full size and adult configuration of the genitalia. Some individuals, however, retained an intermediate appearance even after giving birth. The category of adult male was assigned to males of apparently full size in body, throat, and beard; it may have taken 2 yr beyond the juvenile stage to achieve fully adult characteristics.

Utilizing some behavioral features on occasion to make assignments is almost unavoidable (e.g., an individual making a deep, dawn howl must be an adult male, or an imperfectly seen individual crawling jerkily over the body of a large monkey must be a recent infant). In particular, discussion of variability in development both of behavior and of size and anatomical features requires a large sample of animals with known birth dates.

The infant and juvenile categories were divided into three approximate sizes and the subadults into two. These categories probably have a crude relation to chronological age and may indeed approximately equally subdivide the large category. Conflicts in age assignment with the same animal were usually between two of the adjacent finer categories and were not considered serious. Chalmers (1968a,b) and Alrich-Blake (1970) provide additional discussion on the problems of censusing arboreal monkeys.

DEMOGRAPHIC RESULTS

In the Hato Masaguaral, howlers live in small groups (Table 1, 1972 census; Table 2, 1970 final counts). In

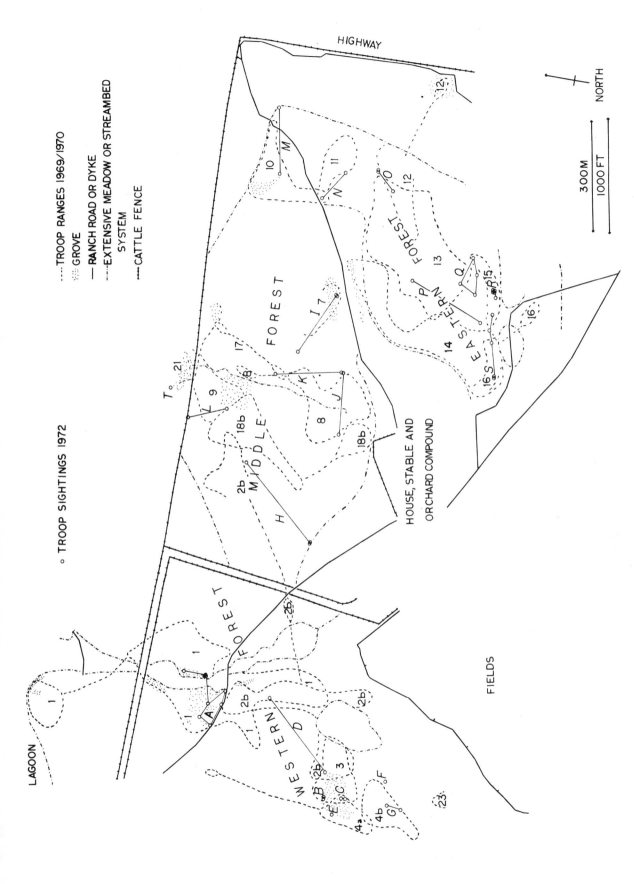

FIGURE 2 Sightings of 1972 troops (lettered) relative to 1969–1970 ranges (numbered). Multiple sightings of the same troop are connected by fine, straight lines. See Table 1 for counts of the 1972 troops and some identities to the 1969–1970 troops presented in Table 2.

103

FIGURE 3 Increase in known population of howlers in Middle and Eastern forests, Hato Masaguaral, between September 11, 1969, and February 25, 1970, as a function of hours worked (searching plus observation). Major jumps in the census are due to the discovery of new troops, while small changes indicate minor changes within troops or improvement of troop counts.

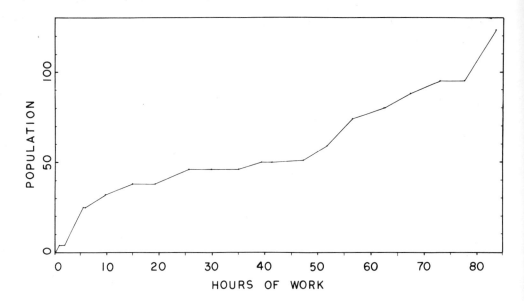

TABLE 1 Summer 1972 Census in Hato Masaguaral: Troops Identified by Letters and Equated, in Some Instances to the Numbered 1969–1970 Troops by Identification of Individuals[a]

Area and Troop	Male Ad.	Subad.	Juv.	Inf.	Female Ad.	Subad.	Juv.	Inf.	Unsexed Inf.	Total
Western Forest										
Troop A (=1)	1		2	2	4		1	1		11
Troop B (=3)	1		1	1	2	1		1		7
Troop C	1				2				1	4
Troop D	1	1	1	1	2		1			7
Troop E	2	2	1		2	1	1	1		10
Troop F	2		1	1	2		2	1		9
Troop G	1	1	1		2	1	1			7
Subtotal	9	4	7	5	16	3	6	4	1	55
Middle Forest										
Troop H	2	1	1	1	3		1	1		10
Troop I (=7)	2		1		2	1	2	1		9
Troop J (=8)	3	1	2		2		1	1		10
Troop K	1			1	2				1	5
Troop L	2	1	2	1	2					8
Troop M	3	1	2	1	2		1			12
Subtotal	13	4	8	5	14	1	5	3	1	54
Eastern Forest										
Troop N	1		1		2		2		1	7
Troop O	2	3	2		4		2	2		15
Troop P	1		1		3	1	1	1		8
Troop Q (=13)	1	1	1	1	3					7
Troop R (=15)	1	1	2		2		1			7
Troop S	2		1	2	3	1	1			10
Subtotal	8	5	8	3	17	2	7	3	1	54
TOTAL	30	13	23	13	47	6	18	10	3	163

[a] Inf. = infant (less than 1 yr); Juv. = juvenile (ca. 1–2.5 yr); Subad. = Subadult (duration ca. 1 yr for females, 1–2 yr for males); Ad. = Adult. See Figure 4 for relation to 1969–1970 ranges.

TABLE 2 Terminal Troop Counts from 1970 Troops Resident in Portions of the Hato Masaguaral Sites Recensused in 1972

Area and Troop	Male				Female				Unsexed Inf.	Total
	Ad.	Subad.	Juv.	Inf.	Ad.	Subad.	Juv.	Inf.		
Western Forest										
1	1				2	1	1	1		6
2b (half)	1	1	1.5		1			0.5		5
3	2	1	1		2		1			7
4a	1		1		3		1	1		7
4b	2		2		3		1			8
23	1				2	1	1			5
Subtotal	8	2	5.5	0	13	2	5	2.5	0	38
Middle Forest										
2b (half)	1	1	1.5		1			0.5		5
7	1		1		2			2		6
8	1	1	3	2	3		2			12
9	2	2		1	2		1			8
10	1	1	2	1	3		1	1		10
17	1				2	1				4
18b	3	1	1		3		1	1		10
Subtotal	10	6	8.5	4	16	1	5	4.5	0	55
Eastern Forest										
11	3	1			2		1	2		9
12	1	1		2	4	1	1	2		12
13	1	2	1	1	2		1	1		9
14	2	1	1		2	1	1	1		9
15	1		1	1	2			1		6
16	2	1			3		1	1		8
Subtotal	10	6	3	4	15	2	5	8	0	53
TOTAL	28	14	17	8	44	5	15	15	0	146

1972, 19 troops were censused west of the major highway crossing the ranch; the average size of these troops was 8.6 ± 2.5. In 1970 the average for 19 troops living in the same area was 7.7 ± 2.3. Approximately 22.5 months separates these counts. Because of the presence of recognizable individuals, some troops are identified in Table 1 as descendants of particular troops in Table 2. Figure 2 shows that such troops tend to be conservative in their ranging tendency. The close correspondence of the number and locations of the other 1972 troops to 1969–1970 ranges is suggestive of other identities. It is assumed that identifiable individuals, usually adult males and females, have not switched troops. Some troops indicated in Neville (1972, Figures 6 and 7) are omitted, either because their ranges fell mostly or totally outside the area recensused in 1972 or because they were judged to have been incursive in the area. Solitary animals and pairs are not indicated in the tables or the figure. To study the probability of individuals switching troops or to census nontroop monkeys would require marking of animals.

Apparently the populations are steady in Eastern and Middle forests and increasing in Western Forest, my impression in 1969–1970 (Neville, 1972). A steady population level need not imply that Middle and Eastern forests are experiencing a death rate equivalent to the birth rate. I suspect that there are various behavioral mechanisms (including the howling vocalizations) that lead to emigration of groups and individuals from favorable but well-populated areas.

Figure 4 presents a comparison of the age–sex structures for 1970 and 1972 in Hato Masaguaral using only the troops that appear in Tables 1 and 2. Rather than compare percentages of the population in each category, actual numbers are used so that overall population change can be observed. The problem of having nonequal age categories is apparent, but the larger juvenile population in 1972 is striking. In Table 3 the finer divisions of juvenile and infant are employed for the 1972 census. Assuming that the small and middle juvenile categories approximate a year of growth and are therefore roughly equivalent in duration to the infant stage, one notes that the second-year

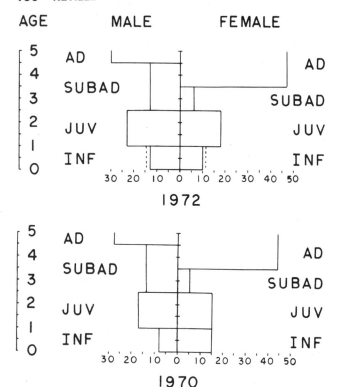

FIGURE 4 Population profiles for the 1972 and 1970 census of Tables 1 and 2. Dotted lines in infant category for 1972 indicate three unsexed infants that were arbitrarily assigned in equal portions to each sex. See text for discussion of the difficulty of correlating age categories with chronological age (present estimate indicated in years).

study (Neville, 1972, Table I). The added value of the latter through the increase in number is partially diminished by the extended time through which the census was taken (average values taken by averaging the first and last representative counts for the troops) and the fact that it contains troops not recensused in 1972. The unusually large juvenile proportion in 1972 is again striking, as well as the general agreement in age proportions between the 1970 census and the larger 1969–1970 census of which the 1970 census is a part.

Comparison of the sex ratios in the 1970 and 1969–1970 censuses indicates that one needs a large sample size for accuracy (in estimation of a hypothetical population equilibrium ratio), the infant ratio in particular being very variable. The adult sex ratios are remarkably stable, part of which is surely accidental. The subadult ratios imply that the subadult male stage lasts at least twice as long as that of subadult females. From knowledge of other monkeys (mostly macaques), one would expect attrition, at least in the form of out-migration, of subadult males also to be higher than that of subadult females. The longevity of howlers in the wild is not known. We are only beginning to appreciate the possibilities for some catarrhine populations (rhesus and Japanese macaques and chimpanzees) on which extended longitudinal studies with identified individuals have been made.

While the population of howler monkeys in this part of Hato Masaguaral is probably growing, the persistence of the trend is untested. Data is almost completely absent on several aspects of howler populations dynamics: the fate and abundance of solitaries, the effects and frequency of group transferals, and the epidemiology of the howler population. A good demographic analysis of a primate population needs a population profile with the age categories stepped in equal progressions (Heltne *et al.*, this volume). Field data are very difficult to convert into this form when the study population lacks a sharp birth season. Many of these questions could be answered if the monkeys could be marked in some fashion that would not

animals number 28 as opposed to 26 first-year animals. One expects some variation from year to year, of course, but one also expects attrition. It appears that the second-year infant group may have been considerably more numerous than the first-year group.

Table 4 compares the proportion and the ratio of females to males in each major age category for the 1972 census, the comparable portion of the 1970 census (roughly 22 months before the 1972 census), and the larger census for the 26 troops of the 1969–1970

TABLE 3   Age Structure of the 1972 Census Using Finer Subdivisions within the Juvenile and Infant Categories

Sex	Adult	Subadult	Juvenile			Infant			Total
			Large	Middle	Small	Large	Middle	Recent	
Male	30	13	8.5[a]	6.5[a]	8	3	9	1	79
Female	47	6	5	8	5	4	6	0	81
TOTAL	77	19	13.5	14.5	13	7	15	4	163[b]

[a] One individual was described in the notes as "large/middlish."

[b] Three infants were not sexed.

TABLE 4   Numbers, Proportion of Total, and Sex Ratio for Each Age Category in the 1972 Census, the Comparable Portion of the Census in 1970, and the Average Count for 1969–1970, Represented by the Larger Sample in Neville (1972, Table 1)

Age Category	1972			1970			1969–1970		
	Number	Percent	Fem./Male	Number	Percent	Fem./Male	Number	Percent	Fem./Male
Adult	77	47.2	1.57	72	49.3	1.57	108	49.2	1.57
Subadult	19	11.7	0.46	19	13.0	0.36	29	13.2	0.31
Juvenile	41	25.1	0.78	32	21.9	0.88	45.5	20.7	1.27
Infant	26	15.9	0.77[a]	23	15.8	1.88	37	16.9	1.23[b]
TOTAL	163	99.9	1.03	146	100.0	1.18	219.5	100.0	1.18

[a] Three infants in the total count were not sexed.

[b] Two and one-half infants were not sexed; the one-half refers to the summarizing procedure wherein the first and last representative counts of a troop were averaged.

preclude future close observer distances and would permit recensusing over several yearly intervals.

CONSERVATION IN VENEZUELA

Venezuela's 91,205,000 ha in 1970 contained approximately 10,399,000 people (*Brittanica Book of the Year, 1972*), signifying that area alone still forms protection for at least some animals. The Venezuelan Division of National Parks has responsibility for more than 1,400,000 ha (i.e., 1.61 percent of Venezuela) apportioned among 10 national parks and 3 natural monuments. Six forest reserves total an additional 1,813,000 ha. The following description has been drawn principally from Acosta-Solis (1968), Beebe (1949), Beebe and Crane (1947), Harrisson (1971), the UN list of National Parks (IUCN, 1967), and informational sheets (undated) from the Venezuelan Ministry of Agriculture and Husbandry, Division of National Parks.

Seven of the national parks are presented in Figure 1 in addition to the locations in which I worked. Several of the northern parks have potential for primate conservation. The most famous of the parks is Henry Pittier (formerly Rancho Grande), which consists of 90,000 ha rising from 0 m on the coast to 2,344 m in the cloud forest of the coastal Andean range north of Maracay. The laboratory buildings at 1,097 m are actively used by research scientists. The park was established in 1937 and contains howlers (personal observation) and capuchins (Beebe, 1949).

The National Park of Guatopo, comprising 92,640 ha ranging from 400 to 2,200 m, was set aside in 1958 to conserve the water supply to Caracas. Conservation of water supply for farming areas and villages has been a prime reason in the formation of a number of the parks, including Henry Pittier; Sierra Nevada de Mérida (160,000 ha ranging from 600 m up to the highest point in Venezuela, Pico Bolívar, at 5,007 m)

(established in 1952); El Avila (66,192 ha to the north of Caracas with elevations from sea level on the coast to 2,700 m) (founded in 1959); and Yurubí in Yaracuy State (founded in 1960 to protect the water supply of San Felipe) (4,000 ha). Yacambú, 9,000 ha in the Sierra Portuguesa in Lara, was founded in 1962. Canaima, in Bolívar, is the only national park south of the Orinoco; this 1,000,000-ha park, founded in 1962, contains the famous Angel Falls.

An eighth national park, the 8,500-ha Cueva de la Quebrada "El Toro" located in the Federal District (established in 1969), includes the cave "El Toro," from which flows an underground river. The cave roosts the same *Steatornis* birds as the Alejandro von Humboldt Natural Monument in Monagas. The 181-ha Humboldt Monument contains the "Caves of the Guácharos," named after its famous nocturnal oil birds (*Steatornis caripensis*). The largest of the three natural monuments is Cerro de María Lionza, 40,000 ha set aside in 1960 in Yaracuy State. Arístides Rojas, 1,630 ha, dates from 1949 and conserves a rock structure near San Juan de los Morros, in northern Guárico.

Harrisson (1971, p. 12) notes that "wildlife refuges" can be developed in the six forest reserves, but "so far no designations are known to have been made. The rainforests of the plains north of the Orinoco (Apure, Barinas, Guárico) need special attention and consideration in this respect." To this, one could add the forests of Lake Maracaibo, Paria Peninsula, the mouth of the Orinoco, and the vast areas south of the Orinoco—these harbor the great majority of the diversity of the Venezuelan primates (Hill, 1960; 1962; Rohl, 1959). Reserves should be set aside in these areas now.

In the llanos, the relatively few concerned landholders provide the major efforts toward active protection of wildlife, e.g., Tomás Blohm on Hato Masagural. Reservoir areas and adjacent protected watershed,

FIGURE 5 Approximate locations of remnant populations of howler monkeys (*Alouatta seniculus insulanus* [Hill, 1960]) and capuchin monkeys (*Cebus albifrons trinitatis* [Hill, 1962]) in Trinidad. Locations provided by Dr. Elisha Tikasingh (personal communication, 1972) from personal knowledge or discussion with others. Map redrawn from Bacon and Ffrench (1972): the three wildlife sanctuaries that report lists as still containing monkeys are shown numbered: (1) Central Range Wildlife Sanctuary, (2) Trinity Hills Wildlife Sanctuary, and (3) Bush Bush Wildlife Sanctuary.

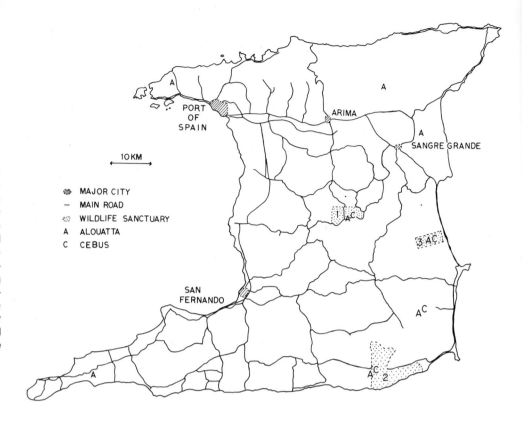

such as the Camatagua Reservoir, could also provide sites for conservation of primates and other wildlife.

CONSERVATION IN TRINIDAD

The fauna of Trinidad is closely related to that of northern Venezuela; Trinidad has probably been separated from the Paria Peninsula only since the end of the Pleistocene era (Vuilleumier, 1971). Figure 5 shows the known locations of remnant populations of howler monkeys (*Alouatta seniculus insulanus*) and capuchin monkeys (*Cebus albifrons trinitatis*) of Trinidad. The distribution is similar to that shown in Downs *et al.* (1955). In general, hunting for food has been intense over the 483,000-ha island, and the expanding human population (945,210 for Trinidad and Tobago, 1970 estimate, *Brittanica Book of the Year, 1972*) has put considerable pressure on forests. The various wildlife sanctuaries of Trinidad and Tobago, most of which either never had or have lost their primate representation, have been described by Bacon and Ffrench (1972). Donald Griffin (personal communication, 1967) reported that monkeys had almost been eradicated from Arima Valley, where the famous William Beebe Tropical Research Station is located. Unreported, isolated populations may remain within the rugged range of hills (heights to 940 m) to the north of the highway between Port of Spain, Arima, and Sangre Grande.

Of the three wildlife sanctuaries in which primates (capuchins and howlers) are reported, the Central Range Wildlife Sanctuary, occupying 2,153 ha, is a part of the Central Range Forest Reserve established in 1922. The sanctuary area came into being in 1934 with the implementation of the 1933 Ordinance No. 35 (Pyke *et al.*, 1972). The Trinity Hills Wildlife Sanctuary occupies 6,483 ha in the Victoria Mayaro Reserve; it is part of the Trinity Hills Reserve, established in 1900. The sanctuary is now under mining lease to Texaco Trinidad Inc. and Trinidad Tesoro Oil Company (Dardaine, 1972). Oil companies may contribute to conservation measures through protection of unused portions of their reserves.

Bush Bush Wildlife Sanctuary, 1,550 ha set aside in 1968, includes the elevated areas of Bush Bush Island and Bois Neuf Forest in the seasonally flooded Nariva Swamp, a major feature of the central eastern side of Trinidad. Bush Bush Forest has been studied since 1959, principally by the Trinidad Regional Virus Laboratory (TRVL). Protection of the area was first carried out by TRVL and then by the New York Zoological Society (NYZS), whose responsibility ended in June 1971 (Tikasingh, 1972). Since that time there has been no systematic protection, and hunters and "fishers" constitute a major threat to the wildlife (Tikasingh, 1972, personal communication).

It was impossible to realistically extrapolate estimates of primate populations in the Bush Bush Wildlife Sanctuary. The population density of 1.1 howlers/ha given in Neville (1972), where the difficulties of this estimate are noted, is surely too low. Only one or two groups of *Cebus* were encountered during the summer of 1968.

The extent and location of wildlife sanctuaries in Trinidad and Tobago in general is adequate and well planned (Tikasingh, personal communication, 1972). However, it is imperative that existing laws be enforced. There is considerable hunting pressure on the animals themselves and a persistent agricultural/ industrial threat to the forests. Trained and responsible forest officials need to be placed in each reserve. Reactivation of research in Bush Bush by an agency such as the TRVL or the NYZS, and a direct approach to enlist the active cooperation of high officials in the oil companies leasing in southern Trinidad, would probably be of great help.

CONCLUSIONS

Our knowledge of the distributions and population parameters of primates in Venezuela and Trinidad is very incomplete. There is little data on the effectiveness of the conservation systems of these two countries in relation to their primate populations. A major problem in Trinidad's Bush Bush Wildlife Sanctuary is enforcing protection. Venezuela has considerable need for the promulgation of more protected areas, as well as enforcement of existing laws.

ACKNOWLEDGMENTS

I wish to particularly thank Tomás Blohm and Elisha Tikasingh for their communications in respect to the conservation systems of Trinidad and Venezuela and for their assistance during my fieldwork. I am also deeply in debt to the many other individuals who have assisted me in my work.

REFERENCES

Acosta-Solis, M. 1968. Protección y conservación de la naturaleza en Sudamérica. Pages 230–250 *in* E. Fittkau, J. Illies, H. Klinge, G. Schwabe, and H. Sioli, eds. Biogeography and ecology in South America, vol. I. Dr. W. Junk Publ., The Hague.

Aldrich-Blake, F. P. G. 1970. Problems of social structure in forest monkeys. Pages 79–101 *in* J. Crook, ed. Social behaviour in birds and mammals. Academic Press, New York.

Bacon, P. R., and R. P. Ffrench, eds. 1972. The wildlife sanctuaries of Trinidad and Tobago. Wildlife Conservation Committee, Ministry of Agriculture, Lands and Fisheries, Trinidad and Tobago. 80 pp.

Beebe, W. 1949. High jungle. Duell, Sloan & Pearce, New York. 379 pp.

Beebe, W., and J. Crane. 1947. Ecology of Rancho Grande, a subtropical cloud forest in northern Venezuela. Zoologica (N.Y.) 32:43–60.

Brittanica book of the year, 1972. Encyclopedia Brittanica, Inc., Chicago.

Chalmers, N. R. 1968a. Group composition, ecology and daily activities of free living mangabeys in Uganda. Folia Primatol. 8(3–4):247–262.

Chalmers, N. R. 1968b. The social behaviour of free living mangabeys in Uganda. Folia Primatol. 8(3–4):263–281.

Chivers, D. J. 1969. On the daily behaviour and spacing of howling monkey groups. Folia Primatol. 10:48–102.

Dardaine, S. 1972. Trinity Hills Wildlife Sanctuary. Pages 25–30 *in* P. R. Bacon and R. P. Ffrench, eds. The wildlife sanctuaries of Trinidad and Tobago. Wildlife Conservation Committee, Trinidad and Tobago.

Downs, E. G., T. H. G. Aitken, and C. R. Anderson. 1955. Activities on the Trinidad Regional Virus Laboratory in 1953 and 1954 with special reference to the yellow fever outbreak in Trinidad, B.W.I. Am. J. Trop. Med. Hyg. 4:837–843.

Harrisson, Barbara. 1971. Conservation of nonhuman primates in 1970. S. Karger, Basel. 99 pp.

Hershkovitz, P. 1949. Mammals of northern Colombia. Preliminary report no. 4: Monkeys (Primates), with taxonomic revisions of some forms. Proc. U.S. Nat. Mus. 98(3232):323–427.

Hershkovitz, P. 1972. Notes on New World monkeys. Int. Zoo Yearb. 12:3–12.

Hill, W. C. O. 1960. Primates: Comparative anatomy and taxonomy. IV. Cebidae, Part A. Edinburgh University Press, Edinburgh. 523 pp.

Hill, W. C. O. 1962. Primates: Comparative anatomy and taxonomy. V. Cebidae, Part B. Edinburgh University Press, Edinburgh. 537 pp.

IUCN (International Union for the Conservation of Nature and Natural Resources). 1967. Liste des Nations Unies des parcs nationaux et réserves analogues. IUCN Publications, New Series, No. 11. Hayez, Louvain, Belgium, 550 pp.

Neville, M. K. 1972. The population structure of red howler monkeys (*Alouatta seniculus*) in Trinidad and Venezuela. Folia Primatol. 17:56–86.

Pyke, S., P. R. Bacon, R. P. Ffrench, and B. S. Ramdal. 1972. Central Range Wildlife Sanctuary. Pages 20–24 *in* P. R. Bacon and R. P. Ffrench, eds. The wildlife sanctuaries of Trinidad and Tobago. Wildlife Conservation Committee, Trinidad and Tobago.

Röhl, E. 1959. Fauna descriptiva de Venezuela (vertebrados), 4th ed. Neuvas Gráficas, Madrid. 516 pp.

Tikasingh, E. S. 1972. Bush Bush Wildlife Sanctuary. Pages 67–71 *in* P. R. Bacon and R. P. Ffrench, eds. The wildlife sanctuaries of Trinidad and Tobago. Wildlife Conservation Committee, Trinidad and Tobago.

Vuilleumier, B. S. 1972. Pleistocene changes in the fauna and flora of South America. Science 173:771–780.

PROBLEMS AND POTENTIALS FOR PRIMATE BIOLOGY AND CONSERVATION IN THE NEW WORLD

Paul G. Heltne *and* **Richard W. Thorington, Jr.**

The major objectives of the conference were to clarify the methodology and evaluate the data necessary for sound conservation and management of wild primate populations. Nearly all the relevant studies of New World monkeys were scrutinized at some point during the proceedings. By detailing the available knowledge of neotropical primates, participants also delineated major gaps in essential information. Thus the papers of the conference address the following important concerns of population biology of New World primates: distributions and absolute and relative abundances; relationship of group density, group size, and group structure of primate species to the total faunal community and the type and quality of forest; and human customs and governmental regulations relevant to persistent threats against primate populations, to primate exports, and to primate conservation. It is quite obvious that both the problems and suggested solutions are broadly applicable to primate conservation on other continents as well. In what follows, we attempt to draw into a cohesive picture the highlights of the formal and informal exchanges and present our synthesis of the information presented at the conference. We will focus the summary around three themes:

1. Crises: The endangered status of the Central American squirrel monkey (*Saimiri sciureus oerstedii*) and the cotton-top tamarin (*Saguinus oedipus*).

2. The limitations of the current data basic to the establishment of sound conservation and management programs: macro-, meso-, and microdistributions; den-

sity estimates; and basic parameters of life history and quality of local habitats.

3. The export trade and other, more severe, pressures on primate populations.

CRISES

The rapid establishment of protected reserves is essential if two species of primates, the cotton-top tamarin and the Central American squirrel monkey, are not to become extinct. The nearly complete destruction of the forests in the ranges of these two species accentuates the "no forests–no primates" principle, which appeared again and again in the conference. This is a dictum that holds rigorously true for all New World primates.

Hernández-Camacho and Cooper map a large species range for *S. oedipus* (see also Hershkovitz, 1949, 1966). This range is documented by museum specimens taken over the last century. However, Hernández-Camacho and Cooper are convinced that this picture is obsolete. Their recent examination of the region shows that primary and second growth forests suitable for cotton-tops have been extirpated from large portions of the historical range. To create grazing land for cattle, forest destruction has proceeded rapidly within the last two decades. During this same period, 30,000–40,000 cotton-tops were exported. The cotton-top has been protected by law in Colombia since 1969, but appropriate habitat reserves are essential to allow the continued natural existence

110

of the tamarin. Patricia Warner is studying one population (deme) of *S. oedipus* in a small remnant forest in northern Colombia. Hopefully, her study will form the basis for the required conservation effort.

Baldwin and Baldwin indicate a similarly extreme habitat depletion for *Saimiri sciureus oerstedii*. Historically, the Central American squirrel monkey was common (Handley, 1966) throughout the forests and woodlands of Chiriqui Province of western Panama. The Baldwins' surveys reveal that most of the Chiriqui forests have already been totally eliminated for pastureland. The continuation of this process is assured by the current Panamanian agrarian reform laws. Returning after a year to recensus a deme of squirrel monkeys, the Baldwins found that their own fairly sizable study plot had been bulldozed. This left stands of trees only in inland and mangrove swamps in that immediate region. It is clear that the governments of Panama and Colombia must act rapidly to designate protected reserves adequate to stave off the extinction of these two species.

The situation appears to be equally critical for several species such as *Leontopithecus rosalia* (see Bridgewater, 1972) and *Saguinus leucopus* (Hernández-Camacho and Cooper, 1976), and only slightly less critical for others. In some species regional populations may be endangered. According to Hernández-Camacho and Cooper this is true for populations of *Aotus trivirgatus*, *Saimiri sciureus*, *Callicebus molloch*, *Cacajao melanocephalus*, *Alouatta seniculus*, *Cebus albifrons*, and *Lagothrix lagotricha*. The Baldwins found *Cebus capucinus* populations more endangered than *Saimiri sciureus oerstedii* in the areas of Chiriqui that they sampled. There was no evidence at all of *Ateles geoffroyi* in the locales the Baldwins visited, though Handley (1966) indicates that it occurred in Chiriqui forests. Freese notes that *Ateles geoffroyi* is extremely rare beyond the boundaries of Santa Rosa Park in the remaining tropical dry forests of Costa Rica. Hunting of *Ateles* and *Lagothrix* for meat and the clearing of the highest and most heterogeneous forests for farm sites were probably straining the reproductive potential of the spider monkey and woolly monkey at La Macarena according to Klein and Klein. In Trinidad the cebus and howler monkeys have been eradicated from all but the most rugged parts of the island and a few small wildlife sanctuaries, themselves unprotected or under mining lease.

The case of *Aotus* is particularly significant, because the night monkey is currently one of the New World primates most in demand for biomedical research. Scientists studying malaria have noted that *Aotus* from northern Colombia are most susceptible to experimental infection with *Plasmodium falciparum*. It is not known whether this susceptibility results from genetic or environmental factors, or both. Because of this susceptibility, *Aotus* are the animals of choice for this research, and approximately 4,000 night monkeys are used in malaria studies each year in the United States. A large portion of these animals come from a relatively small area in the vicinity of Magangue, Colombia, as documented by Green. Conservation of this important population of *Aotus* is a complex problem. Capture of the animals probably occurs as forests are being cleared for other purposes. Thus, destruction of the forests is again a most important aspect of the pressure on night monkeys.

However, in order to assess the effect of habitat change on an *Aotus* population, we need much more information on *Aotus* biology, particularly on the density of animals normally found in different forest types. Unfortunately, most of the field data on the biology of *Aotus* is fragmentary and essentially anecdotal. Hernández-Camacho and Cooper, summarizing what little is known of the habitat requirements of *Aotus*, point out the very hopeful fact that with the exception of mangrove swamps, *Aotus* typically inhabit every major forest zone of Colombia, including second growth and even well-shaded coffee plantations. If complete clearing were avoided, the biomedical supply could be assured with appropriate management applied to small forests on steep or rocky ground, fence rows, streamside tracts of trees, and domestic plantations.

Thorington, Muckenhirn, and Montgomery presented a pilot study of home range and activity patterns of the night monkey. Theirs was a pioneering use of radio-tracking on primates. With this technical advance, they were able to determine that their "study" animal moved short distances during feeding sessions in the vicinity of the home tree. Longer exploratory excursions took it up to 250 m from the home tree, roughly retracing its original pathway on its return journey. Daylight hours were spent at 10–15 m above ground. In the night its travels brought it quite close (3 m) to the forest floor, but most of its activity occurred in the canopy. On large or small branches, it moved quietly and carefully, making few long jumps and using its tail extensively for balance. The animal showed postdusk and predawn activity periods with 1 or 2 hours of low activity in between. Seventy-two percent of its time was spent in an 800 m² area and 85 percent in an area of 0.5 ha. Other night monkeys may have been present in this area as well.

It is not difficult to perceive serious threats to many other populations and species. The assaults on populations of *Lagothrix* are discussed below. For several genera only a dearth of information is available. For example, the notes of Hernández-Camacho and Cooper and the paper by Moynihan constitute almost the total available information on the ecology and natural behavior of the pygmy marmoset (*Cebuella pygmaea*). These authors document the unusual sap-licking habit of the pygmy marmoset. This component of the *Cebuella* diet is in addition to fruit, buds, and insects. The species will come to the ground for insects, its sap holes are low on the trees, and it prefers to make crossings from tree to tree at a low level rather than be exposed to the dangers of predation in the canopy.

Much less is known about the habitats of *Chiropotes*, *Pithecia*, or *Cacajao*. Indeed, *Cebuella*, *Chiropotes*, or *Callimico* may provide the crucial experimental interface for cancer or schistosomiasis, just as *Aotus* is so unexpectedly doing for malaria research. In most cases we lack the detailed information required for intelligent evaluation of the condition of a primate species. We must know where the species lives, both in general and very specific terms, and we must be able to assess the status of populations and the quality of habitats.

DATA BASIC TO PRIMATE CONSERVATION

Distributions

Distributional data are of primary importance in primate conservation. Three sorts of distributions were presented at the symposium: (a) macrodistributions (the presence or absence of a species in a large region of a country or between two major tributaries of the Amazon), (b) mesodistributions (presence or absence of a species in a locale), and (c) microdistributions (distribution of animals within forest types of a particular locale including the study of habitat preferences and resource utilization of the primates). Several papers added abundance, density, biomass, and ethological details to accounts of microdistributions. Ultimately much more distributional data at all three levels will be needed for informed decisions on resource management.

Macrodistribution

Taxonomic and zoogeographic studies of the ceboids are essential to a clear understanding of the problems of primate conservation in the New World. Hernández-Camacho and Cooper draw together for the first time a zoogeography of the primates of Colombia. A summary appears in Table 1. Previously, macrodistributions for Colombian primates were widely scattered through many, often obscure, sources

or, at best, were arranged by genus and again difficult to collate. As they point out, many of the bibliographic sources are old, and Hernández-Camacho and Cooper updated this information with many observations made recently in the field. An inadequately known distribution can make conservation problems seem much less grim than they are and cause us to misdirect our limited remedial efforts. The most critical problems are found among species with limited distributions, usually in northern and central Colombia. In these areas, forest tracts once rich in primates remain only as scattered plots or hedgerows, as has been discussed above. On the other hand Hernández-Camacho and Cooper indicate an enormous range extension for *Callimico goeldii*. This primate, apparently never existing in high densities anywhere in its range, was previously documented only from small pockets in Peru and Brazil.

For several important genera, large regions remain unknown, e.g., the whole area between the Yari River and Apaporis River for *Saguinus* (two species meet in this area), *Pithecia* and *Cacajao* (which apparently replace each other in this region), and *Cebus albifrons*. In the range of *C. albifrons*, large areas are indicated as uncertain; this is also true in the distributions of *Saimiri*, *Aotus*, *Lagothrix*, and *Ateles*. Given the present importance of primates, question marks on distribution maps cry for the necessary field studies to fill out the range or indicate that indeed a species definitely does not exist (or no longer exists) in a region. It may be noted here that vast regions of Brazil, Paraguay, Bolivia, Ecuador, the Guianas, Nicaragua, El Salvador, the Honduras, Guatemala and the Yucatan are unknown so far as the present distribution of primates is concerned. For purposes of conservation and biomedical research supply, it is increasingly important that these gaps be filled.

Important problems of competition, ecological replacement, and speciation are raised by the distribution maps of Hernández-Camacho and Cooper. For example, they indicate a rather narrow zone of contact between *Aotus trivirgatus lemurinus* and *A. t. trivirgatus*. The recent indications (Brumback *et al.*, 1971) of chromosome polymorphism within *A. t. griseimembra* and cytogenetic differences between *A. t. griseimembra* and *A. t. trivirgatus* make the geographically intermediate and ecologically different *A. t. lemurinus* very interesting biologically.

Hernández-Camacho and Cooper also show that the macrodistributions of numerous primate species are coextensive. This may be illustrated by the following example, though similar situations occur elsewhere. In the Amazonas and Putumayo comesarías of Colombia, between the Caqueta River and the Putumayo River, 13 species are found sympatrically: *Cebuella pyg-*

maea, Saguinus nigricollis or *Saguinus graellsi, Saguinus fuscicollis, Callimico goeldii, Saimiri sciureus, Aotus t. trivirgatus, Callicebus torquatus medemi, Pithecia monachus, Alouatta seniculus, Cebus albifrons unicolor, Cebus apella, Lagothrix l. lagotricha,* and *Ateles belzebuth.* How these species divide space and resources among themselves and the arboreal marsupials, rodents, carnivores, birds, and reptiles is one of the truly exciting questions of contemporary primatology.

Mesodistributions

Utilizing a limited number of records, zoogeographers infer presence of a species in the intervening or surrounding territory in the absence of major ecological barriers or changes, e.g., high mountain ranges, large rivers, or the change from forest to open grassland. Field research cannot sample such large areas uniformly, nor are these regions in any sense ecologically uniform. The next step toward sound conservation is to determine the presence or absence of a species in a given district or locale, such as the area of a proposed national park or within the drainage basin of a small river system.

Freese offers such a mesodistribution for the primates of Santa Rosa National Park, an area of northwestern Costa Rica. Previously, few records documented the existence of *Ateles geoffroyi* and *Cebus capucinus* in this region.

Klein and Klein give similar data for La Macarena National Park and several other sites in the area south of Villavincencio in Colombia. They find more species exist on the south bank of the Guayabero River than on the north bank. *Callicebus moloch* is the only "titi" on the northern bank. South of the Guayabero *C. moloch* has a very limited habitat, and *C. torquatus* is dominant. Also present on the south bank, but not found north of the river, were *Cacajao melanocephalus, Cebus albifrons,* and *Lagothrix lagotricha.* Neville indicates the probable distribution of howlers in the national park system in Venezuela.

Mesodistributions of heavily disturbed regions are given by Neville for Trinidad and by the Baldwins for the Chiriqui Province of western Panama. In Trinidad cebus and howlers are restricted to very few areas and may be inadequately protected or actively hunted even there. In Chiriqui the rugged Burica Peninsula nevertheless shows fairly extensive cutting, though many moderately large forest tracts remain containing one or more of the species present—cebus, howlers, and squirrel monkeys. In the remainder of the province, forests exist only as small pockets in swampy lowlands or in larger marshland tracts along the seacoast. *Ateles* and *Aotus* were found in none of the sites visited in western Chiriqui.

Microdistributions (Habitat evaluation)

Information gathering at the microdistributional level is arduous, time-consuming, and confronted with problems of standardization. Uniformity has not been attained in defining forest types or levels of degradation. Strict comparability may not be possible over any great area.

Freese shows that *Alouatta* are largely restricted to limited stands of evergreen trees in the tropical dry forest of Santa Rosa National Park, in Guanacaste Province, Costa Rica. *Ateles* and *Cebus* have a broader distribution within the park. *Ateles* ranges freely in the semideciduous areas, as well as in evergreen patches, throughout the year. *Cebus* are found in evergreen, semideciduous, deciduous, and mangrove stands during all seasons and are not reluctant to make crossings on the ground. Klein and Klein differentiated eight tree communities within the tropical wet forest in La Macarena Park, in Colombia. These communities varied in usefulness, sometimes seasonally, to the monkeys in the Kleins' study area. Forested islands in heavily farmed Chiriqui often do not contain any primates according to Baldwin and Baldwin. *Saimiri* were more numerous than *Cebus* or *Alouatta* in the most disrupted inland swampy areas. *Cebus* are able to utilize mangrove stands, though *Alouatta* and *Saimiri* spend very little time in mangrove areas and are apparently unable to survive in mangrove and marshy scrub communities. Since such plots will soon be the only available primate habitats in many areas, this report takes on tremendous importance.

Obviously, it is impossible to evaluate every area of forest in Central and South America. Thus, methods of extrapolation and estimation become important. Primate distribution and abundance per habitat type must be complemented with detailed information about distribution and abundance of habitat types. Habitat diversity and condition may be assessed at a qualitative level locally. Over broader areas, newer quantitative methods are becoming available. Locally one need not actually see hunters or woodcutters in the forest to determine their presence. Tree species of economic importance or game species such as peccaries, tapirs, and guans will indicate that incursions by man are not yet severe. Trees such as *Cecropia* are typical of the early stages of forest regeneration and are thus a good indication of disturbed habitats. As suggested by the Kleins, key tree species such as *Brosimum* and *Ficus,* which are important in the diets of many primates, may provide evidence of a favorable habitat.

TABLE 1 Summary of Distribution Data in Hernández-Camacho and Cooper

	Amazonas	Putumayo	Caquetá	Vaupés	Guainía	Meta	Vichada	Nariño	Cauca	Valle	Chocó	Córdoba	Huila	Tolima	Quindío	Cundinamarca	Risaralda	Caldas	Antioquia	Boyacá	Santander	Arauca	Bolívar	Sucre	Atlántico	Magdalena	César	Guajira	Norte de Santander
CALLITRICHIDAE																													
Cebuella pygmaea	✓	✓	✓																										
Saguinus nigricollis	Wb	✓	✓																										
Saguinus graellsi		✓																											
Saguinus fuscicollis subsp.	✓		✓	W		SW			SEt																				
Saguinus inustus				✓	✓																								
Saguinus geoffroyi											✓																		
Saguinus oedipus																			N				NW						
Saguinus leucopus														Nt				E	N+E				S						
CALLIMICONIDAE																													
Callimico goeldii	Wh	✓																											
CEBIDAE																													
Aotus t. trivirgatus	✓	✓	✓	✓	✓	S+W	S	SEt	SEt				Eb			E				C+Nt		W							
Aotus t. griseimembra									E			ET							E	W	✓		S	✓	✓	✓	✓	SW	✓
Aotus t. zonalis										W	✓	✓							NW					Wb					
Aotus t. lemurinus								E	C+E	E	Eb		NW,S+E	C+E	✓	C+E	✓	✓	C	WC	E	Wt							
Callicebus moloch ornatus						WC										SE													
Callicebus moloch discolor		SW						SE																					
Callicebus torquatus medemi	SEt	SWt		SE		SW	SE																						
Callicebus torquatus lugens	✓	✓	✓	✓	✓	S+W	S																						S
Saimiri sciureus	✓	✓	✓	✓	✓	S+W	S	SEt	SEt				SC			Eb				C+Nt	W								
Cebus c. capucinus								W	W+Nb	C+Wb	✓	✓					W+C	St	N+W			N	N	N+C					
Cebus albifrons malitiosus																										NE		W	

114

TABLE 1 (Continued)

	Norte de Santander	Guajira	César	Magdalena	Atlántico	Sucre	Bolívar	Arauca	Santander	Boyacá	Antioquia	Caldas	Risaralda	Cundinamarca	Quindío	Tolima	Huila	Córdoba	Chocó	Valle	Cauca	Nariño	Vichada	Meta	Guainía	Vaupés	Caquetá	Putumayo	Amazonas
Cebus a. cesare		SW	N	√																									
Cebus a. versicolor	N+Wt	S	E+S			E	S		√	W	NE+E	E		NW		N													
Cebus a. albifrons																							NE						
Cebus a. unicolor																										E			SWt
Cebus a. yuracus	SEt							W		Nt																			
Cebus a. subsp.	SEt							√		Eh+Nt				Et			C+S				Et,SEt	Eb					√	√	√
Cebus apella																							√		√	√	√	√	√
Pithecia monachus																								SW	√	√			
Cacajao melanocephalus																									√	√			
Alouatta palliata											NW								√	W	W	W							
Alouatta seniculus		SW						√	√	SC		√	√	√	√	√	√	√	Eb	E	E		St		√	√	√	√	√
Ateles paniscus belzebuth	N,Wt, St	St	√	S	√	√	C+S	W	W	Nt+Wt	E+C	E		N		√	√	√	√	SEt	SEt	W		W+S					
Ateles p. hybridus							NW				W+NC		W						√	W	W	W							
Ateles p. robustus																			NW										
Ateles p. grisescens								W												Et,SEt	Et,SEt			Wh					
Lagothrix lagotricha lugens	SEt							W	W	Nt+C				S+SE		S	√										W	W	
Lagothrix l. lagotricha																							Sb	Sb	√	√	√	√	√
Lagothrix l. subsp.							SC											SE											

LEGEND: S, E, N, W, C = southern, eastern, northern, western, central portions; SE = southeastern portion; S + E = southern and eastern portions; b = border or edge; t = tip; h = half; √ = whole province.

On a broader scale, agricultural and forestry reports from primate-producing nations are often valuable, but seldom contain the detail necessary for exacting estimation. Technology is now available for quantitative habitat evaluations over large areas. The basic forest data is supplied by ERTS satellite photos that provide extremely detailed knowledge of the distribution of forest and some differentiation of forest quality. The satellite photos also have the tremendous advantages of frequent update and the possibility of automated analysis. However, no scientific group is organized or funded to extrapolate primate abundance from forest distribution, even if more of the necessary baseline correlations were available.

Density and Biomass

The density and biomass data presented at this conference (summarized in Table 2) can be duplicated for other species of New World primates in only a very few instances: *Callicebus molloch ornatus* (Mason, 1968) and *Saimiri sciureus* (Thorington, 1968) in small forest plots on Hacienda Barbiscal in the llanos of Colombia; Isla Santa Sofia in the Amazon River near Leticia, Colombia, for *Saimiri sciureus* (Bailey *et al.*, 1974); and Warner's study of *Saguinus oedipus* in northern Colombia. Only for Barro Colorado Island (BCI) is there anything approaching time-trend data over a prolonged period.

Biomass estimates, calculated from information in the various papers in this symposium, are amazingly similar within genera. Despite smaller group size, *Alouatta seniculus* troops show the same general biomass relationships as do BCI and Costa Rican howlers. This makes the intense population at Hacienda Barqueta even more phenomenal and supports the Baldwins' interpretation of their study site as a refuge for troops originally dispersed over a much

TABLE 2 Population Density and Biomass.[a]

Species, Locale, and Source	Troop Size	Density of Troops	Range of Troops	No. Individuals per Unit Area	Biomass[b] (kg/ha)	Remarks
Alouatta palliata						
BCI 1932 (Carpenter, 1964)	4–35	23/3,840 acres (1.5/km²)	44–76 ha[c]	398/3,840 acres (26/km²)	1.4	
BCI 1933 (Carpenter, 1964)	4–29	28/3,840 acres (1.8/km²)		490/3,840 acres (32/km²)	1.7	
BCI 1951 (Collias and Southwick, 1951)	2–17	30/3,840 acres (1.9/km²)	12.3–16.2 ha[c]	240/3,840 acres (15/km²)	0.8	
BCI 1959 (Carpenter, 1964)	3–45	44/3,840 acres (2.8/km²)		814/3,840 acres (52/km²)	2.9	
BCI 1967 (Chivers, 1969)	11–18	63–87/3,840 acres (4.0–5.6/km²)	7.9–11.6 ha[c]	0.60–0.80/ha (60–80/km²)	3.3–4.4	
Santa Rosa (Freese)	3–24	8–10/4 km² (2.0–2.5/km²)		70–100/4 km² (18–25/km²)	1.0–1.4	Howlers are found chiefly in evergreen forest at Santa Rosa
Hacienda Barqueta (Baldwin and Baldwin)	7–28	11/20 ha (55/km²)	3.2–6.9 ha	at least 157/20 ha (785–1,050/km²)	45	Density 12–29 times as high as BCI with normal fertility and no indication of food shortage
Burica Peninsula (Baldwin and Baldwin)	10–30	several seen or heard at once				
Inland lowland forests of Chiriqui, Panama (Baldwin and Baldwin)	small	scattered				
Alouatta seniculus						
La Macarena, Colombia (Klein and Klein)	3–6	6–15/780 ha (0.8–2.0/km²)		30–75/260 ha (12–29/km²)	0.6–1.6	
Hato Masagural Venezuela (Neville, 1972 census)	4–15	19/190.6 ha (10.0/km²)	0.66–7.08 ha (aver. 3.21 ha, 1969 census)	16.3/190.6 ha (86/km²)	4.7	
Ateles belzebuth						
La Macarena	17–22 independently locomoting animals	3/780 ha (0.4/km²)	259–388 ha	30–40/260 ha (12–15/km²)	0.6–0.8	

TABLE 2 (Continued)

Species, Locale, and Source	Troop Size	Density of Troops	Range of Troops	No. Individuals per Unit Area	Biomass[a] (kg/ha)	Remarks
Ateles geoffroyi						
Santa Rosa	1–20			110–160/17 km² (6–9/km²)	0.3–0.5	
Cebus capucinus						
Santa Rosa	15–20	15–20/49 km² (0.3–0.4/km²)	0.5 km² (50 ha)	250–350/49 km² (5–7/km²)	0.1–0.2	
Hacienda Barqueta	27–30	1/32–40 ha (2.5–3.1/km²)	32–40 ha	0.7–0.9/ha (70–90/km²)	1.8	Forage and travel 50%–70% of time
Burica Peninsula	20 or more					Males threaten observers
Inland lowland forest of Chiriqui	2–5					Animals flee silently
BCI (Oppenheimer, 1968)	15 or less		0.9 km² (90 ha)			Males threaten observers
C. apella						
La Macarena	6–12 independently locomoting animals	4–6/780 ha (0.5–0.8/km²)		15–25/260 ha (6–10/km²)	0.2	
Saimiri sciureus oerstedii						
Hacienda Barqueta	23–27	2 at 20 ha site (5/km²)	17.5–40 ha	14.7/10 ha, 50/20 ha (147–250/km²)	0.7	Forage and travel 95% of time. Biomass: 1.37–2.32 using 147–250 animals/km²
Burica Peninsula	15–30					
Inland lowland forest of Chiriqui	10–20	1/forest pocket	0.8–2 ha	10–13/ha (1,000–1,300/km²)	9.3–12.1	Forest sizes range from 0.8–2.0 ha
S. sciureus						
La Macarena	25–35 independently locomoting animals	3–6/780 ha (0.4–0.8/km²)		50–80/260 ha (19–31/km²)	0.2–0.3	

[a] Nature of sample in each case can be obtained from Table 4.

[b] Biomass calculations, except for Baldwin and Baldwin, are based on the following figures from Eisenberg and Thorington (1973): *Alouatta*, 5.5 kg; *Ateles*, 5.0 kg; *Cebus*, 2.6 kg; *Saimiri*, 0.93 kg.

[c] From Neville, 1972.

greater area. *Ateles* and *Cebus* from Colombia and Costa Rica show quite similar density, but *Cebus* from Hacienda Barqueta have a 10-times-higher density. *Saimiri* at Hacienda Barqueta and La Macarena are certainly of the same magnitude of biomass, but *Saimiri* stranded in relict inland forest plots have a density 10–20 times as great. Indeed, squirrel monkey biomass in these small forest pockets exceed the biomass of all other species and locales except for the Hacienda Barqueta howlers.

When density estimates can be combined with habitat distinctions, the area over which a forest type remains constant (i.e., similar in species composition, in hunting or logging pressure, or the presence of other arboreal animals) can be used to estimate the total population of a species in a region. Without such a dual data foundation, management has a very unstable basis for decisions regarding permissible levels of harvesting. The comparative figures, produced here for the first time, are tremendously valuable. The consistency within genera suggests the possibility of fairly effective extrapolation. Neville correctly emphasizes the necessity of a large sample size for estimating primate population parameters for a particular habitat type. Even with extensive data on one site, as Neville points out, extrapolation has decreased certainty the farther it goes from the site of census. There probably exists some optimal distance for each species at which a spot survey should be made to check whether extrapolated density estimates are acceptably accurate.

Once baseline densities for habitats of various types and degrees of disturbance are obtained, technically advanced methods such as the ERTS satellite photos may speed estimates for broader areas. With these photos experimental attempts at correlations of forest type with primate density are now under way in Colombia under the joint auspices of INDERENA, PAHO, and ILAR. The primate density estimates produced by this technically advanced system must also be locally verified with spot surveys or "ground truth stations."

To summarize, our ability to estimate whether there are 10 monkeys per hectare or one monkey per 10 hectares for any species or area in the New World tropics remains limited and often uncertain, though the comparative data published in this symposium are certainly a major advance. This lack of sound scientific information explains why no nation has yet established a husbandry-management program for primates akin to that for game animals and birds. It is indeed possible that some areas would sustain a thinning of the primate population and even benefit by limited harvest with appropriate techniques (though we are unaware of such locales). Other areas most assuredly cannot.

Life Tables

As Heltne, Turner, and Scott indicate, careful examination of the age structure of a troop or population is necessary to estimate reproductive success and overall status. Density estimates alone are not a sufficient base for an effective program of conservation. Life tables of wild primates need to be constructed to insure that a healthy population structure can be recognized and maintained. This requires considerable basic knowledge concerning the biology of a particular species, i.e., length of the preadult period, number of years a female can be expected to live after attaining reproductive maturity, and how many infants per year she will produce under normal circumstances. At the moment, data for constructing a life table are being collected for the howler monkeys of BCI by Thorington and co-workers. In order to do so, they must capture, handle, and mark animals. Techniques for doing this have become well developed and almost completely safe. Attempts to gather such data have not even been initiated for other species in other areas.

Age criteria for field evaluation of population structure are extremely important. It was felt by most at the conference that more effort should be made to refine our techniques for estimating the age of wild primates. It would be very valuable to compile a photographic handbook of animals of known age so that fieldworkers could refer to a visual gauge of the changes in physiognomy of animals as they mature. More data are desirable on the eruption and wear of teeth, on the rates of skeletal maturation, and on the weights of animals of known age. Much of this information can be obtained in the field only when the investigator captures and handles his animals. Valuable collaboration in building these references and obtaining the life table

values listed above could come from zoo and institutional breeding programs.

While short-term field studies cannot produce data adequate to establish life tables, such research can produce age–sex class censuses. Repeated censuses can give some indication of the population structure found in normal or stressed situations. Even this approach is technically difficult. Neville correctly emphasized the problems of obtaining accurate data on such basic parameters as total numbers, ages, and sexes of animals in a troop and the necessity of a large sample size for estimating these parameters for a particular habitat type. Certain problems of censusing appear to be greatly diminished if two or three individuals are engaged in the counting process at once (Thorington, Heltne, and Schön, personal observation).

In the absence of clear-cut demographic data, field censuses have not had a theoretical baseline for comparison. Heltne, Turner, and Scott derive an estimate of the status of a troop or population of *Alouatta palliata* by assigning values to a set of demographic parameters (values can easily be adjusted for other species). These authors suggest that one gauge of success in a population is the number of adult females relative to the number of juveniles plus infants (F:J + I ratio). While a dense population can arise from several causes in a disturbed environment, this ratio reflects the degree of reproductive success achieved over a period at least as long as the age to maturity. A minimal and tentative criterion for judging the status of a population or troop of howler monkeys is that there should be at least as many juveniles and infants as adult females. However, for many troops of rural rhesus monkeys, population stability is achieved only when the number of infants and juveniles equals or slightly exceeds the total number of adults of both sexes (Southwick and Siddiqi, 1968). In addition, female rhesus monkeys are thought to have a much longer life span than females of most or all of the New World species of primates, though age to sexual maturity is about the same for rhesus and cebids. Thus, among the howlers, and by extension other cebids, an F:J + I ratio of 1:1 may be an extremely conservative estimate for indicating that a population or a particular troop is adequately replacing itself. Only age-specific birth and death rates for a particular species in a particular habitat can securely supply the value of the replacement level of the F:J + I ratio.

This symposium contributes significant new census data for several species. These figures are gathered and developed further in Tables 3 and 4 and compared with information from BCI. Neville's data (Table 3)

TABLE 3 **Comparison of Three Censuses of *Alouatta seniculus* in a Comparable Area of Hato Masaguaral, Guarico, Venezuela**[a]

Year and Forest Area	Group Size		Composition (%)									M:F (1:—)	F:I (1:—)	F:J + I (1:—)	F:S + J + I (1:—)	One-Male Troops (%)	Sub-adult (%)
	Range	Mean ± SD	M	SM	JM	IM	F	SF	JF	IF	I?						
1969–1970	4–14	8.5 ± 2.5	19	10	9	8	30	3	12	9		1.57	0.56	1.25	1.69	54	51
1970																	
Western	5–10	6.9 ± 2.4	21	5	14	0	34	5	13	7		1.62	0.19	1.00	1.31	55	45
Middle	4–12	8.5 ± 2.4	18	11	15	7	29	2	9	8		1.60	0.53	1.38	1.81	62	53
Eastern	6–12	8.8 ± 1.9	19	11	6	8	28	4	9	15		1.50	0.80	1.33	1.87	50	53
TOTAL	4–12	7.9 ± 2.3	19	10	12	5	30	3	10	10		1.57	0.52	1.25	1.68	55	51
1972																	
Western	4–11	7.9 ± 2.3	16	7	13	9	29	5	11	7	2	1.78	0.63	1.44	1.88	71	55
Middle	5–12	9.0 ± 2.4	24	7	15	9	26	2	9	6	2	1.08	0.64	1.57	1.93	17	50
Eastern	7–15	9.0 ± 3.2	15	9	15	6	31	4	13	6	2	2.12	0.41	1.29	1.71	67	54
TOTAL	4–15	8.6 ± 2.6	18	8	14	8	29	4	11	6	2	1.57	0.55	1.43	1.83	53	53

LEGEND: M = male; F = female; S = subadult; J = juvenile; I = infant.
[a] Data from Neville, 1972, and 1976.

from undisturbed troops of *Alouatta seniculus* show strong consistency from year to year in total numbers per forest and in the proportionality of the various age–sex classes. Especially notable are the very high F:J + I and F:S + J + I ratios. These figures are almost twice as high as comparable values from *A. seniculus* in La Macarena and from *A. palliata* from BCI and Costa Rica. Of interest also is the high proportion of one-male troops in the marginal dry forest environment of Santa Rosa, Costa Rica. Santa Rosa *Ateles* and *Cebus* seem in a depressed state compared to La Macarena *Ateles* and BCI *Cebus* population structures. It is clear that both field research and theoretical developments in the study of population structure deserve to be carried forward to provide a strong foundation for conservation and for investigations into primate ethology.

PRESSURES ON PRIMATE POPULATIONS

Although little is known about the population biology of New World primates, humans inflict three extraordinary pressures on these species: We capture them for pets or experimentation. We hunt them for food. We destroy their habitats over broad areas.

Export Trade

Happily, the wasteful and often inhumane pet trade in monkeys has largely ended. But this relaxation is balanced by the growing research endeavors in the USA, USSR, Japan, Western Europe, and elsewhere, which create an accelerating demand for primates as experimental animals. The testing of plastics, cosmetics, and pharmaceuticals for toxicity and the standardization of vaccines requires many more monkeys than basic biomedical research. Intensive efforts to find preventatives and cures for cancer, malaria, onchocerciasis, and so forth can only increase scientific demands for primate exports. For the first time, Green elucidates in detail one export chain from first sale by the captor to the flight from Barranquilla, Colombia, to the USA. The stresses to the animals en route suggest that even if export for experimentation were the only pressure on the populations, some alteration in the export system will be required if resource management is to be effective. The capture of primates for use in research may be accomplished by several means, with differing effects on wild populations. The least-harmful techniques, often haphazard, are those that result in the removal of only a small portion of juvenile monkeys without endangering future survival of the population. In other cases, the trapping is more efficient and more animals are removed, including some of the breeding adults. Or worst of all, adult females may be shot and their young captured, a practice obviously deleterious to a population. Hunting practices should be better documented and hunting regulations of different countries should be compiled and their varying effects on primate populations assessed.

Primates as Protein

It is now becoming apparent that far more severe pressures are placed on many primate species through hunting for food and through destruction of habitats

TABLE 4   Population Structure

Species, Site, and Source	Nature of Sample	Group Size		Composition ($\bar{x}\% \pm$ SE)				M:F (1:—)	F:I (1:—)	F:J + I (1:—)	One-Male Troops (%)	Sub-adults (%)
		Range	Mean ± SD	M	F	J	I					
Alouatta palliata												
BCI, various authors (see Heltne *et al.*)	167 troops, in 7 censuses, 1932–1972	2–45	15.6 ± 3.7	18 ± 1	43 ± 3	21 ± 2	18 ± 1	2.53	0.42	0.93	23	39
Taboga, Guanacaste, C. R. (Heltne *et al.*)	102 troops, in 9 censuses, 1966–1971	2–39	11.5 ± 2.1	21 ± 2	48 ± 2	21 ± 3	10 ± 2	2.38	0.21	0.65	29	30
LaPacifica, Guanacaste, C. R. (Heltne *et al.*)	10 troops, in 3 censuses, 1966–1970	5–27	11.9 ± 1.7	21 ± 6	47 ± 6	20 ± 2	13 ± 3	2.76	0.29	0.72	44	33
Santa Rosa, Guanacaste, C. R. (Freese)	8 troops sampled Oct. 1971 to April 1972	3–24	8.1 ± 7.1	20	44	24	12	2.23	0.26	0.85	63	36
Hacienda Barqueta, Chiriqui, Panama (Baldwins)	8 troops, Dec. 1970 to Feb. 1971	7–28	18.9 ± 6.6	20	42	20	17	2.06	0.41	0.88	0	37
Burica Penn., Chiriqui, Panama (Baldwins)	Survey, 1968–1970	10–30	—	—	—	—	—	—	—	—	—	—
Inland, lowland forests, Chiriqui, Panama (Baldwins)	Survey, 1968 and 1970	small	—	—	—	—	—	—	—	—	—	—
Alouatta seniculus												
La Macarena, Meta, Col. (Kleins)	6–15 troops, 1967 and 1968	3–6	5 (median)	28	40	22	10	1.43	0.25	0.80	75	32
Hato Masaguaral, Guarico, Venezuela (Neville)	63 troops, in 3 censuses, 1969–1972	4–15	8.4 ± 2.4 (1970, 1972)	19 ± 0.3	30 ± 0.3	35 ± 1	16 ± 1 (J includes subadult males and females)	1.57	0.54	1.73	54	51

TABLE 4 (Continued)

Species, Site, and Source	Nature of Sample	Group Size Range	Group Size Mean ± SD	Composition (X̄% ± SE) M	F	J	I	M:F (1:—)	F:I (1:—)	F:J + I (1:—)	One-Male Troops (%)	Sub-adults (%)
Ateles belzebuth La Macarena, Col. (Kleins)	2 troops, 1967 and 1968	17–22 independently locomoting animals		17	49	17	17	2.9	0.35	0.70	0	52
A. geoffroyi Santa Rosa, C. R. (Freese)	22 troops, Oct. 1971 to April 1972	1–20	—	30	45	14	11	1.5	0.24	0.55	—	33
Cebus capucinus Santa Rosa, C. R. (Freese)	Several troops, Oct. 1971 to April 1972	15–20	—	28	39	22	11	1.4	0.29	0.86	—	33
Hacienda Barqueta, Panama (Baldwins)	1 troop, Dec. 1970 to Feb. 1971	27–30	—	—	—	—	—	—	—	—	—	—
Burica Peninsula, Panama (Baldwins)	Survey, 1968–1970	20 or more	—	—	—	—	—	—	—	—	—	—
Inland, lowland forests, Panama (Baldwins)	Survey, 1968 and 1970	2–5	—	—	—	—	—	—	—	—	—	—
BCI (Oppenheimer, 1968)	10 troops, March 1966 to August 1967	15 or less	—	14 (from typical count, 2M:3F:7J:2I)	21	50	14	1.5–1.9	0.63–0.66	1.80–3.05	30–70	54–64
C. apella La Macarena, Col. (Kleins)	4–6 troops, 1967 and 1968	6–12 independently locomoting animals		—	—	—	—	—	—	—	—	—
Saimiri sciureus oerstedii Hacienda Barqueta, Panama (Baldwins)	2 troops, Dec. 1970–Feb. 1971	23–27	—	10 (J consists of 32J + 10 subadult males)	26	42	22	2.6	0.85	2.08	0	64
Burica Peninsula, Panama (Baldwins)	Survey, 1968–1970	15–30	—	—	—	—	—	—	—	—	—	—
Inland lowland forests, Panama (Baldwins)	Survey, 1968, 1970	10–20	—	—	—	—	—	—	—	—	—	—
S. s. sciureus La Macarena, Col. (Kleins)	3–6 troops, 1967 and 1968	25–35 independently locomoting animals		—	—	—	—	—	—	—	—	—

121

than through the export trade. It is at this point that human politics, economics, and nutrition are extremely relevant. Work on these topics, beyond the scope of this symposium, may be more crucial to the status of New World primate populations than studies of primates themselves. In some parts of Latin America, there is an immense and growing human consumption of monkeys as a major source of dietary protein. Pierret and Dourojeanni (1966) and Mel Neville and co-workers (reports to ILAR Committee on Conservation of Nonhuman Primates) show that each year large numbers of monkeys of several species appear on the Iquitos food market. For example, a minimum of 2,500 dried woolly monkey carcasses are sold annually for food in Iquitos (Neville, personal communication). Woolly monkey meat is also prized in many areas of Colombia according to Hernández-Camacho and Cooper. Many of the monkeys taken for human consumption are eaten along the route to the major market centers. Therefore, we cannot account for the total number of monkeys consumed for meat. It is clear even from the minimal estimates available that this predation pressure is significant, and regionally it may be devastating to some species. In 1973, Peru effectively banned the sale of monkey meat in Iquitos and other major urban markets. Hunting monkey meat for one's family is still permitted and occurs widely.

Agriculture

The most serious threat to primates is the spread of both swidden and mechanized agriculture through the tropical forest in response to the food demands of the exploding local populations or the demands for beef for export to the developed countries. As farms are established, the forest is destroyed and with it the habitat for primates. Monkeys such as *Lagothrix*, which are restricted to primary forests, are eliminated immediately. Like some Old World species such as rhesus and talapoins, a few neotropical primates can make use of human crops. If fruit trees or small plots of banana replace the original forest, *Cebus, Ateles, Saimiri*, and possibly other species may fare quite well. *Cebus* will also consume maize. *Cebus* is considered a crop pest in Colombia (Hernández-Camacho and Cooper) and *Saimiri* and *Cebus* are hunted as pests in Panama (Baldwin and Baldwin). Moynihan believes that *Cebuella* may live, almost as a commensal of man, in small plots or hedgerows and fence lines, provided these contain sufficient sap-producing species. This raises the possibility that monkeys could be trapped for export from these pest or semidomestic populations. This is a possibility to be pursued along with the reservation of sizeable areas of natural

habitat. Even though some New World species may make crossings from tree to tree on the ground, in the last analysis complete destruction of the forest is devastating for New World monkeys.

Clearing of forests is frequently the direct result of a national development or reform policy. Even small tracts of forestland are by definition "unproductive" under the agrarian reform laws of many Latin American countries. "Unproductive" land is taxed so heavily that most owners are forced to cut or bulldoze it, or yield the property in lieu of taxes to the government, which distributes it as part of the agrarian reform program. Land-hungry people promptly slash and burn the forest and plant.

OLD WORLD PRIMATES

We recognize that none of these problems are unique to the conservation of New World monkeys. In Asia and Africa, deforestation proceeds at a rapid pace, even in the absence of warfare. Southwick (personal communication) indicates that in Malaysia the meat of the leaf monkeys (*Presbytis obscurus, P. melalophus*, and *P. cristatus*) is a great delicacy among the aborigines and Chinese. Monkey carcasses are commonly sold in the meat markets in West Africa. In many areas primates may be exterminated because they are pests on human food crops. Nor is the knowledge of population parameters much more advanced for the Old World primate species. Density estimates are available for small areas of Malaysia (Chivers, 1973; Southwick and Cadigan, 1972), Indonesia (Rodman, 1973, Wilson and Wilson, in prep.), Gaboon (Gautier-Hion, 1973), Cameroon (Gartlan and Struhsaker, 1972), and for various locales for baboons (DeVore and Hall, 1965). Except for surveys near Delhi (Southwick and Siddiqi, 1968; Mukherjee and Mukherjee, 1972), such information is not available for the major supply areas of rhesus monkey export. Density estimates for *Presbytis entellus* from two areas in India appear in Sugiyana (1964). A summary of available population density estimates, not necessarily related to microdistributional parameters, is supplied by Jolly (1972).

CONCLUSION

The reduction of primate populations and their biological support systems has fortunately found recognition in the emerging conservation movements in many Latin American countries. While striving to set up adequate reserves for their uniquely rich flora and fauna, these nations are stringently limiting exports to prevent, if possible, the total loss of their splendid national heritage. However, the cost of establishing, policing, and managing reserves is high. The cost of research outlined above is certainly significant. Money is also required to educate legislators to value stands of trees and to teach rural people to harvest monkeys in such a way as to secure a sustained crop.

One possible source of funds would be a very high export tax on each primate leaving its country of origin. Attempts to breed rhesus monkeys in this country are producing animals costing between $500 and $1,000 each. This is several times the cost of a wild-caught animal delivered to the laboratory. Much of the cost of a wild-caught animal stems not from the value of the monkey as it leaves India but rather from expenses due to shipping, short-term quarantine, and minimal veterinary medical care in this country. Clearly, the Indian government could charge $300 or $400 per monkey and still export less-costly animals than those bred outside of India. No doubt an analogous situation pertains to the costs of breeding New World primates. (It is clear that the tax must be assessed at the point of departure, or the possibility of irresistible wealth for the rural woodsman would quickly lead to the total extinction of all wild primates.) In theory, the export tax could then help pay for the education of legislators and rural people, the purchase of reserves, and the general management of primate resources. It is hoped that countries possessing wild primates will recognize the extreme value of these assets, control them appropriately, and charge accordingly.

Certain analogies between the crises in primate supply and oil supply will be apparent to most by this point. However, the microdistribution and abundance of primates has not been assessed with nearly the attention given to those parameters for oil. Oil is, of course, a nonrenewable commodity, while primates are a renewable resource. However, this ability of primates to reproduce themselves is highly dependent on some very important habitat requirements. Primates are not only being harvested, but also the very ecological foundations for their capabilities of renewing themselves for further harvest are being destroyed.

Institutional users of primates must act now to press for the soundest possible basis for management and control of natural monkey populations. This will often require what some like to call alpha-level biological research. The technical and scientific dividends that such studies can now pay are absolutely essential to continued investigation in other areas that utilize primates as experimental animals. Both contractual and grant arrangements should be facilitated to bring the knowledge of the distribution, abundance, and status of primate resources up to the level of our knowledge

of oil reserves. Possibly a United Nations conference is necessary to set up an international agenda for primate conservation and management. In terms of potential benefits to humans, the national and international value of primates as a biomedical resource deserves appropriate accounting. Even were it not so, the species and the places where they live are beautiful and precious in their own right.

REFERENCES

Bailey, R. C., R. S. Baker, D. S. Brown, P. von Hildebrand, R. A. Mittermeier, L. E. Sponsel, and K. E. Wolf. 1974. Progress of breeding project for nonhuman primates in Colombia. Nature 248:453–455.

Bridgewater, D. D., ed. 1972. Saving the lion marmoset. The Wild Animal Propagation Trust, Wheeling, W. Va.

Brumback, R. A., R. D. Stanton, S. A. Benjamin and C. M. Lang. 1971. The chromosomes of *Aotus trivirgatus* Humboldt 1812. Folia Primatol. 15:264–273.

Carpenter, C. R. 1964. Behavior in red spider monkeys in Panama. Pages 93–105 *in* C. R. Carpenter, ed. Naturalistic behavior of nonhuman primates. The Pennsylvania State University Press, University Park.

Chivers, D. J. 1969. On the daily behavior and spacing of the howling monkey groups. Folia Primatol. 10:48–102.

Chivers, D. J. 1973. An introduction to the socio-ecology of Malayan forest primates. Pages 101–146 *in* R. P. Michael and J. H. Crook, eds. Comparative ecology and behavior of primates. Academic Press, New York.

Collias, N. E., and C. H. Southwick. 1951. A field study of population density and social organization in howling monkeys. Proc. Am. Philos. Soc. 96:143–156.

DeVore, I., and K. R. L. Hall. 1965. Baboon ecology. Pages 20–52 *in* I. DeVore, ed. Primate behavior. Holt, Rinehart and Winston, New York.

Eisenberg, J. F., and R. W. Thorington, Jr. 1973. A preliminary analysis of a neotropical mammalian fauna. Biotropica 5:150–161.

Gartlan, J. S., and T. T. Struhsaker. 1972. Polyspecific associations and niche separation of rain-forest anthropoids in Cameroon, West Africa. J. Zool. 168:221–266.

Gautier-Hion, A. 1973. Social and ecological features of talapoin monkey-comparisons with sympatric cercopithecines. Pages 147–170 *in* R. P. Michael and J. H. Crook, eds. Comparative ecology and behavior of primates. Academic Press, New York.

Handley, C. O., Jr. 1966. Checklist of the mammals of Panama. Pages 753–795 *in* R. L. Wenzel and V. J. Tipton, eds. Ectoparasites of Panama. Field Museum of Natural History, Chicago.

Hernández-Camacho, J., and R. W. Cooper. 1976. The nonhuman primates of Colombia. This volume.

Hershkovitz, P. 1949. Mammals of northern Colombia. Preliminary report number 4: Monkeys (Primates), with taxonomic revisions of some forms. Proc. U.S. Nat. Mus. 98:323–427.

Hershkovitz, P. 1966. Taxonomic notes on tamarins, genus *Saguinus* (Callithricidae, Primates), with descriptions of four new forms. Folia Primatol. 4:381–395.

Jolly, A. 1972. The evolution of primate behavior. The MacMillan Company, New York.

Mason, W. A. 1968. Use of space by *Callicebus* groups. Pages 200–216 *in* P. C. Jay, ed. Primates. Holt, Rinehart and Winston, New York.

Mukherjee, R. P., and G. D. Mukherjee. 1972. Group composition and population density of rhesus monkey [*Macaca mulatta* (Zimmerman)] in northern India. Primates 13:65–70.

Oppenheimer, J. R. 1968. Behavior and ecology of the white-faced monkey, *Cebus capucinus*, on Barro Colorado Island, Canal Zone. Ph.D. Thesis. University of Illinois, Urbana.

Pierret, P. V., and M. J. Dourojeanni. 1966. La caza y la alimentación humana en las riberas del río Pachitea, Perú. Turrialba 16:271–277.

Rodman, P. S. 1973. Population composition and adaptive organization among orangutans of the Kutai Preserve. Pages 171–209 *in* R. P. Michael and J. H. Crook, eds. Comparative ecology and behavior of primates. Academic Press, New York.

Southwick, C. H., and F. C. Cadigan, Jr. 1972. Population studies of Malaysian primates. Primates 13:1–18.

Southwick, C. H., and M. R. Siddiqi. 1968. Population trends of rhesus monkeys in villages and towns of northern India, 1959–1965. J. Anim. Ecol. 37:199–204.

Sugiyama, Y. 1964. Group composition, population density and some sociological observations of Hanuman langurs (*Presbytis entellus*). Primates 5:7–37.

Thorington, R. W., Jr. 1968. Observations of squirrel monkeys in a Colombian forest. Pages 69–85 *in* L. A. Rosenblum and R. W. Cooper, eds. The squirrel monkey. Academic Press, New York.

SUBJECT INDEX

AUTHOR INDEX

PARTICIPANTS

Mrs. Janice Baldwin, Santa Barbara, California

Dr. John Baldwin, Department of Sociology, University of California, Santa Barbara

Dr. Irwin Bernstein, Department of Psychology, University of Georgia, Athens

Mrs. Susan Bernstein, Athens, Georgia

Dr. C. Ray Carpenter, Department of Psychology, University of Georgia, Athens (deceased)

Dr. Robert Cooper, División de Parques Nacionales y Vida Silvestre, INDERENA, Bogotá, Colombia, and Smithsonian–Peace Corps Environmental Program, Colombia/Peace Corps Conservation Program (Present address: Department of Zoology, California State University, San Diego)

Mr. Lawrence Dresdale, Department of Psychology, University of Georgia, Athens (Present address: Department of Psychology, Bucknell University, Lewisburg, Pennsylvania)

Dr. Norris Durham, Department of Anthropology, Wayne State University, Detroit, Michigan

Mr. Curtis Freese, Peace Corps, Servicio de Parques Nacionales, Costa Rica. (Present address: Department of Pathobiology, The Johns Hopkins University School of Hygiene and Public Health, Baltimore, Maryland)

Mr. Ken Green, Peace Corps Conservation Program, División de Parques Nacionales y Vida Silvestre, INDERENA, Bogotá, Colombia (Present address: National Zoological Park, Smithsonian Institution, Washington, D.C.)

Dr. Paul Heltne, Department of Anatomy, The Johns Hopkins University School of Medicine, Baltimore, Maryland

Dr. Jorge Hernández-Camacho, Wildlife Subprogram, División de Parques Nacionales y Vida Silvestre, INDERENA, Bogotá, Colombia

Mr. Michael Kavanagh, Department of Psychology, University of Georgia, Athens (Present address: School of Biological Sciences, University of Sussex, Brighton, England)

Mrs. Dorothy Klein (Present address: Edmonton, Alberta, Canada)

Dr. Lewis Klein, Department of Anthropology, University of Illinois, Urbana (Present address: Department of Zoology, University of Alberta, Edmonton, Alberta, Canada)

Dr. Martin Moynihan, Smithsonian Tropical Research Institute, Box 2072, Balboa, Canal Zone

Dr. Nancy Muckenhirn, National Zoological Park, Smithsonian Institution, Washington, D.C. (Present address: ILAR, National Academy of Sciences, Washington, D.C.)

Dr. Melvin Neville, Department of Anthropology, University of California, Davis

Mr. Tom Patterson, Department of Psychology, University of Georgia, Athens (Present address: Department of Psychology, University of California, Riverside)

Dr. Anita Schwaier, Battelle-Institut, Frankfurt, Federal Republic of Germany

Dr. Richard W. Thorington, Jr., Primate Biology Program, Smithsonian Institution, Washington, D.C.

CONTRIBUTORS

Dr. G. G. Montgomery, Smithsonian Tropical Research Institute, Box 2072, Balboa, Canal Zone

Dr. Norman J. Scott, Jr., Department of Biological Sciences, University of Connecticut, Storrs (Present address: Department of Biology, University of New Mexico, Albuquerque)

Dr. Dennis C. Turner, Department of Mental Hygiene, The Johns Hopkins University School of Hygiene and Public Health, Baltimore, Maryland (Present address: Zoologisches Institut and Museum der Universitat, Zurich, Switzerland)

Neotropical primates : field studies and conservation : proceedings of a symposium on the distribution and abundance of neotropical primates, / R. W. Thorington, Jr. and P. G. Heltne, editors ; . Committee on Conservation of Nonhuman Primates, Institute of Laboratory Animal Resources, Assembly of Life Sciences, National Research Council. — Washington : National Academy of Sciences, 1976.

v, 135 p. : ill. ; 28 cm.

"Sponsored by the Battelle Seattle Research Center and the Institute of Laboratory Animal Resources, National Research Council."
Includes bibliographies and indexes.

329197

(Continued on next card)

76-10786
MARC

76

Neotropical primates ... 1976.   (Card 2)

ISBN 0-309-02442-0

1. Primates—Congresses.   2. Mammal populations—Latin America—Congresses.   I. Thorington, Richard W.   II. Heltne, P. G.   III. Battelle Memorial Institute, Columbus, Ohio.   Seattle Research Center.   IV. National Research Council.   Institute of Laboratory Animal Resources.   V. National Research council.   Institute of Laboratory Animal Resources.   Committee on Conservation of Nonhuman Primates.

QL737.P9N38   599'.8'098   76-10786
MARC

Library of Congress   76